Aeronautical Radio Communication Systems and Networks

Aeronautical Radio Communication Systems and Networks

Dale Stacey

John Wiley & Sons, Ltd.

Other Wiley Editorial Offices

John Wiley & Sons Inc., 111 River Street, Hoboken, NJ 07030, USA

Jossey-Bass, 989 Market Street, San Francisco, CA 94103-1741, USA

Wiley-VCH Verlag GmbH, Boschstr. 12, D-69469 Weinheim, Germany

John Wiley & Sons Australia Ltd, 42 McDougall Street, Milton, Queensland 4064, Australia

John Wiley & Sons (Asia) Pte Ltd, 2 Clementi Loop #02-01, Jin Xing Distripark, Singapore 129809

John Wiley & Sons Canada Ltd, 6045 Freemont Blvd, Mississauga, ONT, Canada L5R 4J3

Wiley also publishes its books in a variety of electronic formats. Some content that appears
in print may not be available in electronic books.

British Library Cataloguing in Publication Data

A catalogue record for this book is available from the British Library

ISBN 978-0-470-01859-0 (HB)

Typeset in 10/12pt Times by TechBooks, New Delhi, India.

Printed and bound in Great Britain by Antony Rowe Ltd, Chippenham, Wiltshire

Contents

Preface   xvii

    Dedications   xviii
    About the Author   xviii
    Revisions, Corrections, Updates, Liability   xix
    Book Layout and Structure   xix

1   Introduction   1
   1.1 The Legacy   1
   1.2 Today and the Second Generation of Equipment   1
   1.3 The Future   3
   1.4 Operational and User Changes   3
   1.5 Radio Spectrum Used by Aviation   4
      1.5.1 Convergence, Spectrum Sharing   6
   1.6 Discussion of the Organizational Structure of Aviation
       Communications Disciplines   6
      1.6.1 International Bodies   7
      1.6.2 Example National Bodies   7
      1.6.3 Industrial Interests   7
      1.6.4 Example Standards Bodies and Professional Engineering Bodies   7
      1.6.5 Users/Operators   8

2   Theory Governing Aeronautical Radio Systems   9
     Summary   9
   2.1 Basic Definitions   10
      2.1.1 Notations and Units   10
   2.2 Propagation Fundamentals   11
      2.2.1 Electromagnetic Vectors   11
      2.2.2 Polarization   11
      2.2.3 Speed of Propagation and Relationship to Wavelength and Frequency   11
   2.3 Power, Amplitudes and the Decibel Scale   14
   2.4 The Isotropic Power Source and Free Space Path Loss   15
      2.4.1 Definition of Isotropic   15
      2.4.2 Derivation of Free Space Path Loss Equation   15

2.4.3	Power Flux Density	17
2.4.4	Electric Field Strength	17
2.4.5	Relationship Between Field Strength and Transmitted Power	18
2.5	Radio Geometry	19
2.5.1	Radio Horizon Calculations	19
2.5.2	Earth Bulge Factor – k Factor	22
2.5.3	Nautical Mile	23
2.5.4	Great-circle Distances	24
2.6	Complex Propagation: Refraction, Absorption, Non-LOS Propagation	25
2.6.1	Refraction	26
2.6.1.1	Layer Refraction	26
2.6.1.2	Obstacle Refraction	26
2.6.2	Attenuation from Atmosphere Absorption	28
2.6.2.1	Water Absorption	28
2.6.2.2	Oxygen Absorption and Other Gases	28
2.6.3	Non-LOS Propagation	30
2.6.3.1	Propagation – Ground Wave	30
2.6.3.2	Reflection and Multipath	30
2.6.3.3	Propagation – Sky Wave	32
2.6.4	Propagation to Satellite	36
2.6.4.1	Propagation Distance	36
2.6.4.2	Atmospheric Losses	36
2.7	Other Propagation Effects	37
2.7.1	The Doppler Effect	37
2.7.1.1	Example	37
2.7.1.2	Answer	38
2.8	Modulation	38
2.8.1	The Modulation Conundrum	40
2.8.2	The Analogue and Digital Domains	40
2.8.3	Amplitude Modulation (AM)	41
2.8.3.1	DSB-AM	41
2.8.3.2	The VHF Aeronautical Mobile Communications (Route) Service (AM(R)S)	43
2.8.3.3	Single Sideband (SSB) Modulation	46
2.8.3.4	The Aeronautical HF System and Other SSB Systems	48
2.8.3.5	Suppressed Carrier Double Side Band AM	48
2.8.4	Frequency Modulation	49
2.8.4.1	Capture Effect (Hysteresis)	49
2.8.5	Digital Modulation	50
2.8.5.1	Amplitude Shift Keying (ASK)	50
2.8.5.2	Amplitude Modulated Minimum Shift Keying (AM–MSK)	51
2.8.5.3	Baud/Bit Rate and 'M-ary' ASK	52
2.8.5.4	Bipolar and Differential	52
2.8.5.5	Frequency Shift Keying	53
2.8.5.6	Phase Shift Keying	53
2.8.5.7	Quadrature Amplitude Modulation (QAM) and Trellis Code Modulation (TCM)	58

       2.8.5.8 Trellis Code Modulation   59
       2.8.5.9 Gaussian Frequency Shift Keying (GFSK)   60
2.9 Shannon's Theory   62
    2.9.1 Non-Errorless Transmission   62
2.10 Multiplexing and Trunking   62
    2.10.1 Frequency Division Multiplexing (FDM)   63
    2.10.2 Trunking   63
       2.10.2.1 Example   63
    2.10.3 Time Division Multiplexing (TDM)   65
    2.10.4 Orthogonal Frequency Division Multiplexing (OFDM) and
          Coded OFDM   65
2.11 Access Schemes   66
    2.11.1 Frequency Division Multiple Access (FDMA)   66
    2.11.2 Time Division Multiple Access (TDMA)   67
    2.11.3 Code Division Multiple Access (CDMA)   67
       2.11.3.1 CDMA Principles   69
       2.11.3.2 Frequency Domain Duplex (FDD) and Time
              Domain Duplex (TDD)   70
       2.11.3.3 CDMA Applications   71
2.12 Mitigation Techniques for Fading and Multipath   71
    2.12.1 Equalization   71
    2.12.2 Forward Error Correction and Cyclic Redundancy Checking   72
    2.12.3 Interleaving   72
    2.12.4 Space Diversity   74
    2.12.5 Frequency Diversity   75
    2.12.6 Passive Receiver Diversity   75
2.13 Bandwidth Normalization   77
2.14 Antenna Gain   80
    2.14.1 Ideal Isotropic Antenna   80
    2.14.2 Practical Realizations   81
    2.14.3 Some Common Antennas Used for Aeronautical Communications   82
       2.14.3.1 The Dipole   82
       2.14.3.2 The Folded Dipole   82
       2.14.3.3 Quarter-Wave Vertical Antenna   82
       2.14.3.4 5/8 λ Vertical Antenna   83
       2.14.3.5 Yagi Antenna   84
       2.14.3.6 Log Periodic Antenna   84
       2.14.3.7 Parabolic Dish Antennas   86
2.15 The Link Budget   87
2.16 Intermodulation   88
    2.16.1 Third-order, Unwanted Harmonics   88
    2.16.2 Higher Order Harmonics   92
2.17 Noise in a Communication System   92
    2.17.1 Thermal Noise   92
    2.17.2 Natural Noise   92
    2.17.3 Man-made Noise and Interference   92
    2.17.4 Sky Noise   93

2.18 Satellite Theory   93
 2.18.1 Extended Noise Equation   93
 2.18.2 *G/T*   93
 2.18.3 The Link Budget Equation   94
 2.18.4 Noise Temperatures   95
  2.18.4.1 Receiver Side of the Reference Point   95
  2.18.4.2 Antenna Side of the Reference Point   95
2.19 Availability and Reliability   99
 2.19.1 Definitions   99
 2.19.2 The Reliability Bathtub Curve   99
 2.19.3 Some Reliability Concepts   100
 2.19.4 Overall Availability of a Multicomponent System   101
  2.19.4.1 Serial Chain   101
  2.19.4.2 Parallel Chain   101
  2.19.4.3 The Reliability Block Diagram   102
 Further Reading   104

3 VHF Communication   105
 Summary   105
3.1 History   105
 3.1.1 The Legacy Pre-1947   105
 3.1.2 1947 to Present, Channelization and Band Splitting 106
  3.1.2.1 Channel Splitting   108
 3.1.3 Today and 8.33 kHz Channelization   108
 3.1.4 Into the Future (Circa 2006 Plus)   109
3.2 DSB-AM Transceiver at a System Level   110
 3.2.1 System Design Features of AM(R)S DSB-AM System 110
  3.2.1.1 Availability and Reliability   113
  3.2.1.2 RF Unbalance   113
  3.2.1.3 System Specification   113
3.3 Dimensioning a Mobile Communications System–The Three Cs 113
 3.3.1 Coverage   115
  3.3.1.1 Voting Networks and Extended Coverage   117
 3.3.2 Capacity   120
 3.3.3 Cwality (Quality)   122
3.4 Regulatory and Licensing Aspects   123
 3.4.1 The Three As   123
  3.4.1.1 Allocation   123
  3.4.1.2 Allotment   124
  3.4.1.3 Assignment   124
  3.4.1.4 Utilization Profile   124
3.5 VHF 'Hardening' and Intermodulation   125
 3.5.1 Receiver Swamping   125
 3.5.2 Intermodulation   126
3.6 The VHF Datalink   126
 3.6.1 Limitations with VHF Voice   126
 3.6.2 The History of Datalink   127

	3.6.3	System-Level Technical Description	128
		3.6.3.1 ACARS/VDL0/VDLA	128
		3.6.3.2 VDL1	129
		3.6.3.3 VDL2	130
		3.6.3.4 VDL Mode 3	134
		3.6.3.5 VDL4	138
	3.6.4	Overview of the Modes – A Comparison	140
	3.6.5	Services over Datalink	140
	3.6.6	Future Data Applications	140
	Further Reading		143

4 Military Communication Systems 145
  Summary 145
4.1 Military VHF Communications – The Legacy 145
4.2 After the Legacy 146
4.3 The Shortfalls of the Military VHF Communication System 147
4.4 The Requirement for a New Tactical Military System 147
4.5 The Birth of JTIDS/MIDS 147
4.6 Technical Definition of JTIDS and MIDS 148
  4.6.1 Channelization 148
  4.6.2 Link 4A Air Interface 148
  4.6.3 Link-11 Air Interface 148
  4.6.4 Link 16 – Air Interface 149
  4.6.5 Access Methods 151
  4.6.6 Link 16 Data Exchange 152
  4.6.7 Jitter 152
  4.6.8 Synchronization 152
  4.6.9 Sychronization Stack 152
    4.6.9.1 Header 153
    4.6.9.2 Data Packing 153
    4.6.9.3 Standard Double Pulse Format 154
    4.6.9.4 Packed 2 Single Pulse Format 154
    4.6.9.5 Packed 2 Double Pulse Format 155
    4.6.9.6 Packed 4 Single Pulse Format 155
  4.6.10 Other Salient Features of JTIDS/MIDS 156
  4.6.11 Overlay with DME Band 156

5 Long-Distance Mobile Communications 157
  Summary 157
5.1 High-Frequency Radio – The Legacy 157
5.2 Allocation and Allotment 158
5.3 HF System Features 158
  5.3.1 Transmitter 159
  5.3.2 Receiver 159
  5.3.3 System Configuration 159
  5.3.4 Selective Calling (SELCAL) 159
  5.3.5 Channel Availability 160

5.4	HF Datalink System	162
	5.4.1 Protocol	162
	5.4.2 Deployment	163
5.5	Applications of Aeronautical HF	163
5.6	Mobile Satellite Communications	165
	5.6.1 Introduction	165
	5.6.1.1 Geostationary Satellite Systems	165
	5.6.1.2 Low-Earth Orbit Satellite Systems	167
	5.6.1.3 Medium-Earth Orbit Satellite System	168
	5.6.2 Geostationary Services System Detail	168
	5.6.2.1 The AMS(R)S Satellite System	168
	5.6.3 Antenna System Specifications	171
	5.6.3.1 Satellite Antenna Figure of Merit (G/T)	172
	5.6.3.2 Antenna Discrimination	172
	5.6.3.3 Rx Thresholds	173
	5.6.3.4 Tx EIRP Limits	174
5.7	Comparison Between VHF, HF, L Band (JTIDS/MIDS) and Satellite Mobile Communications	175
5.8	Aeronautical Passenger Communications	175
	Further Reading	175
6	**Aeronautical Telemetry Systems**	177
	Summary	177
6.1	Introduction – The Legacy	177
6.2	Existing Systems	178
	6.2.1 A Typical Telemetry System Layout	179
	6.2.1.1 Transmitter Side (On-board Aircraft Components)	180
	6.2.1.2 Receiver Side (High-performance Ground Station)	181
	6.2.1.3 On-board System Duplication and Ground Backhaul Infrastructure	181
	6.2.2 Telecontrol	182
6.3	Productivity and Applications	182
6.4	Proposed Airbus Future Telemetry System	183
	6.4.1 Channelization Plan	183
	6.4.2 System Components	183
	6.4.3 Telemetry Downlink	183
	6.4.4 Telecommand Uplink	184
6.5	Unmanned Aerial Vehicles	185
7	**Terrestrial Backhaul and the Aeronautical Telecommunications Network**	187
	Summary	187
7.1	Introduction	187
7.2	Types of Point-to-point Bearers	188
	7.2.1 Copper Cables	188
	7.2.2 Frequency Division Multiplex Stacks	189
	7.2.3 Newer Digital Connections and the Pulse Code Modulation	189
	7.2.4 Synchronous Digital Hierarchy, Asynchronous Transfer Mode and Internet Protocol	191

7.2.5	Fibre Optic		191
7.2.6	Private Networks and the Aeronautical Telecommunications Networks		192
7.2.7	PTT-Offered Services		194
7.2.8	Radio Links		194
	7.2.8.1	Fixed Radio Link Design	194
7.2.9	VSAT Networks		197
	7.2.9.1	VSAT Radio Link Budget	197
7.2.10	Hybrid Network		199

8 Future Aeronautical Mobile Communication Systems 201

	Summary		201
8.1	Introduction		202
8.2	Near-term Certainties		202
	8.2.1	Universal Access Transceiver	202
		8.2.1.1 Frame Structure	202
		8.2.1.2 UAT Transceiver Specification	203
		8.2.1.3 UAT Modes of Operation	204
		8.2.1.4 Message Types	204
		8.2.1.5 Application and Limitation of UAT	205
		8.2.1.6 Further Reading	205
	8.2.2	Mode S Extended Squitter	205
		8.2.2.1 Mode S Introduction	205
		8.2.2.2 Pulse Interrogations and Replies	206
		8.2.2.3 Further Reading	207
	8.2.3	802.xx Family	207
		8.2.3.1 802.16	208
		8.2.3.2 Specification	209
		8.2.3.3 Application and Limitations	210
8.3	Longer Term Options		210
	8.3.1	Analysis	210
	8.3.2	Answer	210
	8.3.3	The Definition Conundrum	211
		8.3.3.1 The Requirements or the Operational Scenario	212
		8.3.3.2 Technology Options and Frequency Band	213
		8.3.3.3 Spectrum Requirements	214
	8.3.4	A Proposal for a CDMA-based Communication System	214
	8.3.5	Software Defined Radio	217
	Further Reading		219

9 The Economics of Radio 221

	Summary		221
9.1	Introduction		221
9.2	Basic Rules of Economics		221
9.3	Analysis and the Break-even Point		222
9.4	The Cost of Money		222
	9.4.1	Some Basic Financial Concepts	223
	9.4.2	Inflation	224

9.5	The Safety Case	225
9.6	Reliability Cost	226
9.7	Macroeconomics	227

10 Ground Installations and Equipment — 229
Summary — 229
10.1 Introduction — 229
    10.1.1 Environment — 229
        10.1.1.1 Indoor Environment — 229
        10.1.1.2 Outdoor Environment — 230
10.2 Practical Equipment VHF Communication Band (118–137 MHz) — 233
    10.2.1 VHF Transmitters — 233
    10.2.2 VHF Receivers — 233
    10.2.3 VHF Transmitter/Receiver Configurations — 235
        10.2.3.1 VHF Single-channel Dual Simplex Station Site
                Configuration — 235
        10.2.3.2 VHF Multichannel, Duplicated Base Station — 236
    10.2.4 VHF Cavity Filters — 236
    10.2.5 VHF Combiner, Multicouplers, Switches and Splitters — 237
    10.2.6 Other Radio Equipment — 238
        10.2.6.1 HF — 238
        10.2.6.2 Microwave Point-to-point Equipment — 240
        10.2.6.3 Satellite Equipment — 240
        10.2.6.4 Voice/Data Termination, Multiplex and Other
                Line-terminating Equipment — 241
        10.2.6.5 Future Communication Equipment — 241
    10.2.7 Peripheral Equipment — 243
        10.2.7.1 Mains/AC Service — 243
        10.2.7.2 DC Supplies — 244
        10.2.7.3 Heating Ventilation, Air Conditioning — 244
        10.2.7.4 Pressurization — 244
10.3 Outdoor — 245
    10.3.1 Transmission Lines (VHF, L Band and Microwave) — 245
    10.3.2 Antenna Engineering — 245
        10.3.2.1 Antenna Location and Application — 245
        10.3.2.2 Antenna Selection — 247
        10.3.2.3 Alignment and Optimization — 248
        10.3.2.4 Practical Antennas — 248
    10.3.3 Towers or Masts — 254
    10.3.4 Equipment Room — 255
    10.3.5 Equipment Racks — 257

11 Avionics — 259
Summary — 259
11.1 Introduction — 259
11.2 Environment — 259

11.2.1 Temperature 261
    11.2.1.1 Outside 261
    11.2.1.2 Interior 262
11.2.2 Pressure 262
    11.2.2.1 External Pressure 262
    11.2.2.2 Internal Pressure 262
11.2.3 Equipment Testing 262
11.2.4 Apparent Wind Speed 262
11.2.5 Humidity: 0–100 % 264
    11.2.5.1 External 264
    11.2.5.2 Internal 264
    11.2.5.3 General 264
11.2.6 RF Environment, Immunity, EMC 268
11.2.7 Environmental Classification 268
11.3 Types of Aircraft 268
11.3.1 Private Aircraft 269
11.3.2 General Aviation 269
11.3.3 Commercial Aviation 270
11.3.4 Military Aviation 271
11.4 Simple Avionics for Private Aviation 272
11.5 The Distributed Avionics Concept 273
11.5.1 Data Bus Standards 273
    11.5.1.1 ARINC 429 Standard 273
    11.5.1.2 ARINC 629 Standard 277
    11.5.1.3 ARINC 659 278
    11.5.1.4 Fibre-distributed Data Interface (FDDI) 278
11.5.2 Power Supply System 279
    11.5.2.1 Power Subsystem on an Aircraft 280
    11.5.2.2 Example The Boeing 777 280
    11.5.2.3 28 V DC 281
    11.5.2.4 Flight Management System Monitoring of Circuit Breakers 281
11.6 Avionic Racking Arrangements 282
11.6.1 ATR and MCU 282
11.6.2 Cooling 283
11.6.3 Back Plane Wiring 283
    11.6.3.1 Index Pin Code 284
11.6.5 Other Standards 284
11.7 Avionic Boxes 284
11.7.1 VHF Transceivers 284
    11.7.1.1 Transmitter Specification 285
    11.7.1.2 Receiver Specification 286
    11.7.1.3 Navigation Communication Control Panel 287
11.7.2 HF Radios 289
    11.7.2.1 Technical Specification 289
    11.7.2.2 Transmitter 289
    11.7.2.3 HF Physical Specification 290
    11.7.2.4 Power 290

11.7.2.5 HF Built-in Test Equipment	291
11.7.2.6 HF Antenna Tuner and Coupler	291
11.7.2.7 Dual System Interlocks	292
11.7.2.8 HF Data Radio	292
11.7.3 Satellite Receiver System Avionics	293
11.7.3.1 Receiver Specification	293
11.7.3.2 Size Specification	293
11.7.4 Other Equipment	294
11.8 Antennas	294
11.8.1 VHF Antennas	294
11.8.1.1 Whip Antennas	295
11.8.1.2 Blade Antennas	297
11.8.1.3 Compound Antennas	298
11.8.2 HF Antennas	298
11.8.2.1 Wireline	298
11.8.2.2 Probe Antennas	299
11.8.2.3 Cap Antennas	300
11.8.2.4 Shunt Antennas	300
11.8.2.5 Notch Antenna	300
11.8.2.6 Antenna Couplers	300
11.8.3 Satellite Antennas	300
11.9 Mastering the Co-site Environment	301
11.10 Data Cables, Power Cables, Special Cables, Coaxial Cables	303
11.11 Certification and Maintaining Airworthiness	303
11.11.1 Certification	303
11.11.2 EUROCAE	304
11.11.3 Master Minimum Equipment List	304
Further Reading	304
12 Interference, Electromagnetic Compatibility, Spectrum Management and Frequency Management	307
Summary	307
12.1 Introduction	308
12.2 Interference	308
12.2.1 Sources of Interference	308
12.2.1.1 Accidental or Inadvertent Interference	309
12.2.1.2 Intended or Purposeful Interference	310
12.2.2 Interference Forms	310
12.2.3 Immunity and Susceptibility	311
12.2.4 Testing for Interference	313
12.3 Electromagnetic Compatibility	314
12.3.1 Analysis	314
12.3.2 Out of Channel, Out of Band, Spurious Emissions	316
12.3.3 EMC Criteria	317
12.3.3.1 Building a Compatibility Matrix	318

12.4   Spectrum Management Process   318
         12.4.1   Co-channel Sharing and Adjacent Channel and Adjacent
                  Band Compatibility   319
         12.4.2   Intrasystem and Intersystem Compatibility   319
         12.4.3   Intrasystem Criteria   320
         12.4.4   Intersystem Criteria   320
                  12.4.4.1 Two Aviation Systems   320
                  12.4.4.2 Two Systems: One of Them Not Aviation Safety
                            of Life   320
         12.4.5   WRC Process and the Review and Amend Cycles   321
12.5   Frequency Management Process   322
         12.5.1   Example   322
         12.5.2   Emergency Frequency (Three-channel Guard Band Either Side)   322
         12.5.3   SAFIRE (Spectrum frequency information repositary)   324
         Further Reading   324

Appendix 1 Summary of All Equations (Constants, Variables
            and Conversions)   325
Appendix 2 List of Symbols and Variables from Equations   333
Appendix 3 List of Constants   335
Appendix 4 Unit Conversions   337
Appendix 5 List of Abbreviations   339

Index   **345**

Preface

You may ask why I wrote this book. There are many, many personal reasons as with any author I suppose. The first two reasons and probably the most important are my love of flying and my love of radio engineering. This may sound rather dull but I love flying in any machine be it balloon, glider, propeller aircraft, microlight through to airline jets and the experience of it. The more I do it the more I feel I understand it.

A relative once asked me, 'how does an aircraft fly?' I thought for a while, of how to try and explain the fundamentals of physics and aerodynamics which I feel privileged to have had a fundamental education in. After further thought I realized how I take it all for granted like the vast majority of the people, and despite this education and the sound engineering principles, I still find that flying defies all our instinct and it truly is difficult to explain.

I also find the whole topic of radio propagation equally magical. Again, how can it work when we cannot see it? How can signals travel through apparent nothingness. How can we predict it? The physical equations are all there to describe it in great detail, however it too defies a layman's logic.

If we now marry these two topics together we get Aeronautical Radio Communications—the discipline. This concept is maybe also hard to grasp for most of us and I include myself in this. Writing this book has been a journey of self-discovery and actually showed to myself how much I do not know about the subject rather than how much I know, but hopefully going through this motion has enabled me to know where to look for information when I do not have it to hand.

On the engineering level, some of the system building blocks described may seem very primitive and out of date, especially the legacy aspects, but on another level they are proven to be effective and reliable and this prerequisite knowledge is a fundamental requirement when moving to the design and implementation of the next generation of equipment. There is also the added dimension of thinking about the users of the systems who have a vital role in defining the architecture.

Over the years I have set about collecting the information basis for how the separate aeronautical and radio systems work and I kept them in a file with all the equations I ever used. With time this has grown and initially I have built courses for radio engineers and aviators alike; however, I always planned to put all this information in one tidy place. This is an attempt to do exactly this. It was always my intention to clean up the notes I had and formalize them somehow—hence this book.

I do not pretend that this book has everything on mobile radio communications in it or everything to do with aeronautical mobile radio; however, hopefully it provides the basis for much of it and some explanation, guidance and direction to where further reading material may be available. It is also not intended to be the end of the subject. This topic is continually growing, adapting and getting updated and I have attempted to capture this in the most up-to-date snapshot.

If you do find discrepancies or changes, I would appreciate any comments or information you equally can share with me. In the interim, I hope it provides you with good background in the knowledge you seek.

You can contact me on dale.stacey@consultacom.com

Happy reading!

Dedications

To my wife Mary and two little angels Caitlin and Isla, thank you for your sacrifices of my family time and your support to write this book.

Thanks also and in not any particular order of gratitude to Barbara d'Amato, Alan Jamieson, Kors van der Boogard, John Mettrop for your initial enthusiastic comments and reviews when commencing this journey. Thanks additionally to those who contributed to the book either directly or indirectly: Liviu Popiescu, Roger Kippenberger, Carol Szabo and Stan Jenkins.

Thanks to all my other aviation and radio colleagues, peers and bosses for sharing your experiences, visions and information and making this possible, notably Norman Rabone, John Franklin, Geoffrey Bailey, Christian Pelmoine and Howard Morris.

Thanks to my University and School Teachers who provided me with the basic education and training for this career path. Particularly Mr Sparrow, Dr Wills, Mr Crawford, Dr Aggarwal, Dr Redfern.

Thanks to my Mother Elizabeth and Father Derek for their help at the start and throughout my life and my brothers Paul and Glen who have been indirectly involved in this project.

About the Author

Since graduating from the University of Bath in the United Kingdom in 1988 with a BSc (Hons) degree in Electrical and Electronic Engineering and becoming a Chartered Engineer in the United Kingdom in 1991 and Australia in 1993, Dale Stacey has worked extensively as a Radio Systems Engineer and Project Manager in many arenas all over the world. For the last 15 plus years of which he has been consulting.

Projects have included feasibility studies, planning and design work, installation and commissioning, project management, operation/maintenance and network management of systems. Technologies have included microwave radio links, VHF/UHF mobile systems, GSM 3G, WiMAX and private mobile systems, VSAT satellite systems.

Assignments have included work with oil companies, utility and PTT companies, mining companies, mobile operators, banks, equipment manufacturers and computer network providers, Internet service providers (ISPs) and federal and local government departments in mainly Australia, Asia, North America and Europe.

More recently projects have concentrated on radio systems used in the aviation industry. The author has consulted and worked with Eurocontrol, ICAO, IATA, various government administrations, air navigation service providers (ANSPs) and aeronautical organizations and companies internationally.

The author has dual Australian/British citizenship and spends his time flying around these continents playing with radios as one would expect.

The author derives a living from his consultancy services and teaching in radio engineering, particularly aeronautical mobile radio. More information on training and consultancy services can be found at www.consultacom.com, or you can send an email to dale.stacey@consultacom.com.

Revisions, Corrections, Updates, Liability

I would strongly appreciate feedback as to the content, correctness and ongoing relevance to each of the sections in this book, topics that need deeper elaboration or new topics that should be incorporated. I promise to read all comments and include them as necessary in any future updates. I do believe that this is the best process for improvement. Substantial contributions on your part will be rewarded with a current or future copy of the book and acknowledgement.

Whilst trying to uphold the greatest professionalism obliged by the professional institutions I believe in and belong to, I have endeavoured to provide accurate and unambiguous information. It is hoped that with review and subsequent editions the material can be continually improved. Your help is appreciated in this process.

Book Layout and Structure

The following chapters are generally laid out in a chronological order so the reader can skip parts depending on their subject knowledge or interest. In addition to this there is a matrix layout separating theory (Section A) and practice (Section C) with an intermediate layer called *system level* (Section B) which bridges the gap between theory and practice describing the various building blocks. Thus to a degree the topics are repeated three times with the emphasis changing from theory, system building blocks to practical realizations, so the reader can go back to first principles at any time or concentrate on the system level or physical realizations.

Where content does not sit logically with any of these main sections, special appendices have been compiled, in particular for a summation of all the formulae, list of variables, list of acronyms, constant and unit conversions, etc.

1 Introduction

1.1 The Legacy

The start of the new millennium marks two special centenaries: 100 years of manned flights since the Wright brothers flew the first ever manned heavier than air flight (a total distance of a few hundred feet in December 1903) and also 100 years since the first successful long-distance radio transmissions by Marconi at the end of the Nineteenth century and for the first time across the Atlantic in 1902.

Both of these inventions have revolutionized the world. In many ways the revolutions have only just begun. In the field of aviation, we have seen Concorde and travel to space in the last 50 years. Flying for leisure, the start of Space Tourism and even proposed intercontinental rocket services are perceivable in the not too distant future. Star Wars is the reality!

Likewise, in radio there are revolutions going on in the field of personal communications, in much more recent times with individual mobile phones being the norm and usually incorporating new advanced data services, TV media and video all in one small unit that slips into the back pocket. This as such has replaced what a whole office typing pool, mainframe computer and broadcasting house once did and the threat is *even more progress*: evolution and even revolution with the next generation of intelligent, cognitive and software radio. This is perceived and technically feasible but still really waiting to happen.

The changes in the aviation industry are arguably more conservative and have been slower than the personal communications revolution. The first radio communications were pioneered in the 1920s with tangible on-board transceivers emerging between the war years and with the main standards and practices in aeronautical VHF communications emerging in the late 1940s. These have, arguably, not significantly changed since then. This has been mainly due to very robust and proven systems (for example, the mainstay VHF communication system is testimony to this) that have served us well and is also due to the airlines' reluctance to undergo the time- and cost-intensive process of re-equipping and change (Figure 1.1).

1.2 Today and the Second Generation of Equipment

Today, there is a requirement to enhance the legacy of mobile communication services to provide the users with more functionality, flexibility, immunity to interference (both RF and

Aeronautical Radio Communication Systems and Networks   D. Stacey
© 2008 John Wiley & Sons, Ltd

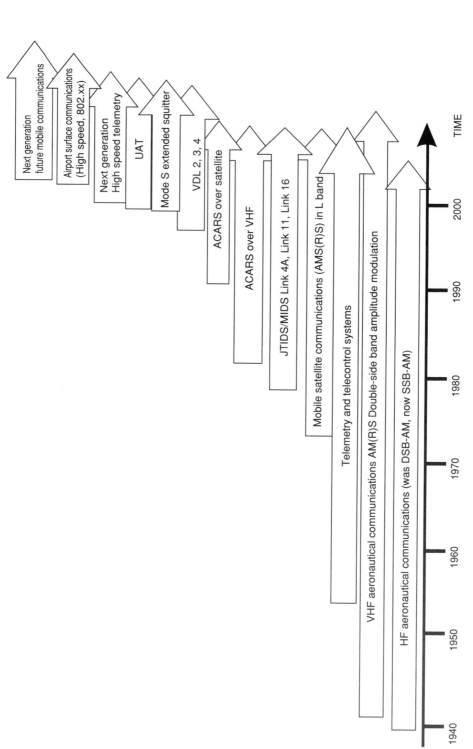

Figure 1.1 Evolution of aeronautical mobile radio systems.

malicious) and reliability. To an extent, this is already well underway by introducing datalink services such as ACARs and VHF datalink and aeronautical satellite services as a second generation stop gap. The 'stop gap' should be emphasized. As with many of these systems, the engineering has been 'shoe-horned' into existing spectrum allocations or using proprietary technologies almost in experimental conditions. Whilst this has bought time, the solutions are not optimized for technology, application and spectrum efficiency and are all the time aging and becoming less relevant.

1.3 The Future

The technology is already ripe for the next (third) generation of communication systems in aviation and the unit cost of this equipment is ever decreasing. The next years will see some decisive changes in aeronautical communications being driven by the availability of this technology and also by the congestion and shortfalls in the legacy systems which are becoming more exaggerated and exasperating every day. Also it is clear that a rationalization of all the systems is required to simplify long term equipage. In contrast, we should not forget our terrestrial mobile communications counterparts (public mobile services) which have already realized much of their third-generation systems and are already planning for fourth- and even fifth-generation systems. Aeronautical communications lag in this deployment but have the advantage to be able to benefit from their experience and even plagiarizing their technology lessons and development work by effectively purchasing plain off-the-shelf modular radio equipment based on these standards. Of course, aviation also has analogous requirements to these other industry sectors transposed to fractionally different scenarios.

1.4 Operational and User Changes

It should never be forgotten that the operational aspects are ever changing, with an emphasis on increased safety statistics, reduced delays for aircraft in all phases from ground turn around, en route and approach stacking, and for greater automation, i.e. less work load on individual air traffic controllers. The user requirements are fast changing from the legacy of system we have from postwar times to fully computerized systems with redundancy provisions.

The customer market profile has totally changed. From the middle of the Nineteenth century and arguably still till the 1980s and 1990s, aircraft transport was historically only available to the upper class and business elite. Today, it very much competes with cars and trains and in some cases has become cheaper than the cost of leaving your car at or getting to the airport. The consequential change in demand has been exponential. In addition, this has changed the airline market profile and drastically the aircraft density; in given air volumes and airports this in turn impacts on operational changes.

The civilian fleets are constantly changing and getting bigger with an emphasis on capacity throughput in high-density sectors – hence, for example, the new Airbus A380. The economic model looks to increase fuel-burn efficiency with litres per passenger mile being the benchmark to improve upon.

There is a growing requirement and commitment to using unmanned arial vehicles (UAVs) in civilian as well as military airspaces, which place a whole new operating concept and requirement on the aeronautical communications systems.

There is a greater need for data interactions between aircraft and ground and for other aircraft to bring in some new navigational and surveillance concepts such as free routes flying (where aircraft adopt a trajectory of least distance akin to great circle routes, instead of the traditional air corridors still used today).

Also, in automatic dependent surveillance (ADS) pilots will attain greater local traffic awareness and responsibility from regularly up-linking adjacent aircraft positions. There is also a strategy to move to greater automated air traffic control, fully computerized with intervention by exception or under conflict only. A new communication system will enable the move to these more efficient operations. This will become critical in the immediate future as fuel prices continue to rise and impact the very fragile economic business cases of the airlines.

1.5   Radio Spectrum Used by Aviation

Figures 1.2 and 1.3 in their broadest senses depict the radio spectrum used by aeronautical communications today.

The subject matter of most interest is probably the VHF communication band, HF band and satellite bands, but the future communications bands should also be stressed, which could likely be VHF (108–137 MHz), L band (960–1215 MHz), S band (2.7–3.1 GHz) and C band (5.000–5.250 GHz) or a hybrid of these. Today, these are only partly defined but will be ratified in the aeronautical agendas planned for the next World Radio Conferences in 2007 and 2011.

Also shown in the figures are adjacent allocations to navigation and surveillance functions and some of the lesser known obscure allocations to specialist services. This figure is generic and applied on a worldwide basis as per the aviation requirement; however, it should be noted that there are some slight regional and individual sovereign state allocation variations that are not discussed here (for a fuller discussion see ITU Radio Regulations, http://www.itu.org).

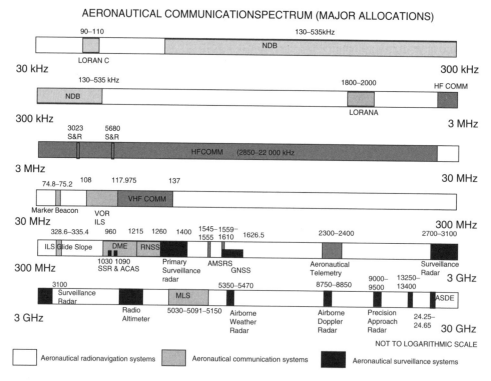

Figure 1.2   Communications radionavigation and surveillance bands.

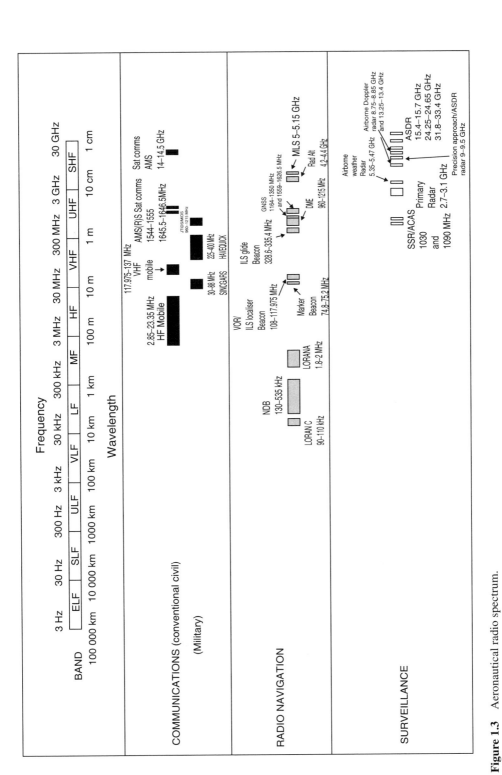

Figure 1.3 Aeronautical radio spectrum.

1.5.1   Convergence, Spectrum Sharing

The concept of convergence is worth mentioning at this stage as well. Historically, separate allocations have been made for the communication, navigation and surveillance functions (sometimes denoted as CNS) for aviation services as defined in ITU. With the spectrum resource being a limited commodity, there has been a growing tendency and impetus towards sharing radio spectra between radio services. This trend is set to continue but also with seeing a merging of these traditional CNS applications to share the same band. These trends somewhat complicate the business of spectrum allocation, sharing and protection from harmful interference. This will be discussed later.

1.6   Discussion of the Organizational Structure of Aviation Communications Disciplines

Finally, by way of an introduction, it is important to mention some of the important stakeholders in the aviation arena (Figure 1.4). Apologies are made in advance if this list is incomplete and it is in no particular order. It is an attempt to capture the relationships.

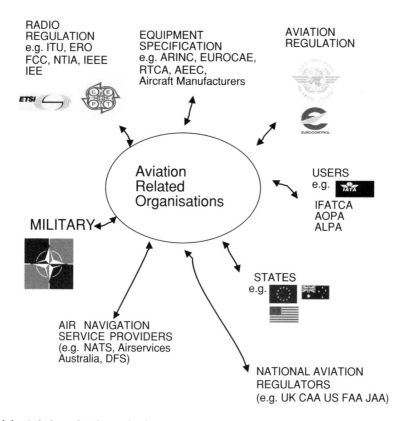

Figure 1.4   Aviation-related organizations.

1.6.1 International Bodies

The International Civil Aviation Organization (ICAO) (see www.icao.int) was formed in December 1944 to provide guidance for setting up standards and recommended practices for the civil airline industry, to promote safety, to help facilitate international air navigation and to harmonize the international regulatory scene.

The International Air Transport Association (IATA) (see www.iata.org) in its own words 'represents, leads and serves the airline industry'; its membership consists of the majority of world airlines. Complete listing of airline membership is on their web page.

The North Atlantic Treaty Organization (NATO), (see www.nato.int) is an international body among other things responsible for harmonizing and organizing the military aspects of aviation in the north Atlantic Europe and America and coordinating with its civilian counterpart (ICAO).

Eurocontrol (see www.eurocontrol.int) is a European wide body responsible for the harmonization and safety of European skies in its 'one sky for Europe' policy.

1.6.2 Example National Bodies

In each country, there are regulatory bodies governing the legal and regulatory aspects of flight within that state. For example, in the United States there is the Federal Aviation Administration (FAA) (see www.faa.gov), in France there is the Direction Générale de l'Aviation Civile (DGAC) (see www.dgac.fr), in the United Kingdom there is the UK Civil Aviation Authority (CAA) (see www.caa.co.uk), and these organizations are generally reflected in each state. The Joint Aviation Authority (JAA) (see www.jaa.org) is partially a European and North American wide representation of the CAA, concentrating on airworthiness, safety aspects and harmonizing of CAA goals.

Also in each sovereign state there is generally an Air Navigation Service Provider; in the United Kingdom, for example, this is National Air Traffic Services (NATS) (see www.nats.co.uk), in Switzerland it is Skyguide (see www.skyguide.ch), in Germany Deutsche Flugsicherung (DFS) (see www.dfs.de).

1.6.3 Industrial Interests

Examples include manufacturers such as Airbus (see www.airbus.com), Boeing (see www.boeing .com), Bombardier (see www.bomabier.com), etc. (Their suppliers and associated aerospace industries are not listed here.)

1.6.4 Example Standards Bodies and Professional Engineering Bodies

There are also a handful of standardizations bodies; some of them of relevance to this book include the following:

- Aeronautical Radio Incorporated (ARINC) (see www.arinc.com);
- European Organisation for Civil Aviation Equipment (EUROCAE) (see www.eurocae.org);
- Radio Technical Commission for Aeronautics (RTCA) (www.rtca.org);
- Airlines Electronic Engineering Committee (AEEC);
- European Telecommunications Standards Institute (ETSI) (www.etsi.org);

- SITA
- European Conference of Postal and Telecommunications Administrations (CEPT) (www.cept.org);
- European Radiocommunications Office (ERO) (see www.ero.dk);
- International Telecommunications Union (ITU) (www.itu.org);
- The Institute of Electrical and Electronic Engineers, Inc. (IEEE) (www.ieee.org);
- The Institution of Engineering and Technology (IET) (www.iee.co.uk).

1.6.5 Users/Operators

As well as IATA already mentioned, some of the other user groups include the following:

- International Federation of Air Traffic Controllers Associations (IFATCA) (see www.ifatca.org);
- Aircraft Owners and Pilots Association (AOPA) (see www.aopa.org), sometimes called General Aviation;
- Airline Pilots Association (ALPA) (see www.alpa.org).

Again, this is only the start of a list and some of the major players.

2 Theory Governing Aeronautical Radio Systems

Summary

This section of the book looks at the physical equations and theory governing radio and aeronautical radio. In particular, it starts from the very basic equation sets and builds up from first principles the derivations of the everyday equations used by most radio engineers and often taken for granted.

Radio waves are often described as having the properties of light and can be described by traditional 'wave' physics. In contrast, sometimes radio waves can be seen to exhibit 'particle' properties or quantum properties. Another important facet of radio engineering is the geometry and the relationship between the transmitters and receivers, and the earth and space through which the radio waves are propagating.

This book is explicitly about radio systems as used in the aeronautical industry; hence this aspect is developed and the derivations are focused in this direction. However, the majority of the equations and derivations described herein are generic and can be applied independently or are beyond the scope of just aeronautical radio systems.

It is important to gather the experience of all the dimensions of the radio physical properties and collect them in one place. That is the purpose of this Chapter 2. It draws on the disciplines of pure physics and electrical engineering and augments them with the mathematics of geometry, reliability, probability, traffic theory and ultimately aeronautical engineering considerations. It may be impossible to collect all the dimensions of radio or the infinite formulae that can be derived from the hundreds of equations already presented here, but hopefully this is a humble beginning.

Examples are offered up to help the reader understand the derivations and formulae at every stage. A full list of all formulae used in this book is summarized in Appendix 1.

Aeronautical Radio Communication Systems and Networks   D. Stacey
© 2008 John Wiley & Sons, Ltd

2.1   Basic Definitions

Maybe the best place to start is with a few basic definitions from which some more complex definitions can be elaborated.

Radio frequency (RF) – The number of cycles per second (or hertz, Hz) of a radio wave.

Transmitter (Tx) – A device that converts electrical signals into emitted RF energy. It is the source of a radiated wave or perturbation.

Receiver (Rx) – A device that receives RF energy and converts it back into electrical signals.

Propagation – The ability of a radio wave to travel between a transmitter and a receiver in free space or in another medium such as air.

Aerial or Antenna – This can be considered as a part of the transmitter or receiver system (or it can be in both). It is usually a passive device that converts electrical signals straight into radio waves.

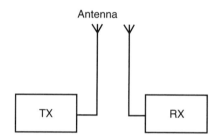

2.1.1   Notations and Units

It is interesting to note that these definitions can be found in many places such as dictionaries, thesauruses, the Institution of Engineering and Technology UK library,[1] the National Physics Laboratory of the UK[2] and the International Telecommunications Union (ITU),[3] to name just a few sources. More curiously, they can be seen to vary very slightly depending on which institution is being considered and the strict application of use and sense. It is important to keep an open mind here so that the discussion can be developed. When transitioning the disciplines from pure physics to electrical engineering to mathematics, the symbols and notation used in formulae can change (a classic example may be the square root of −1, which in physics or engineering is usually denoted by j and in mathematics as i). For certain disciplines and applications a very strict definition is required, particularly at interface points. In this book every attempt has been made to use the standard notation used by electrical engineering disciplines and the ITU. It is suggested that the reader consult Appendix 2, where each of the symbols and variables is listed, and Appendix 3, where each of the physical constants used is listed.

Also with regard to units, it is usual in modern physics and electrical engineering to standardize on the SI unit of measurement. For example, the metre is used for height and distance. This sometimes may conflict with aeronautical industry convention, which is to use feet for altitude (and flight levels, which is feet with the last two zeros knocked off) and nautical miles (NM) for range distance. This confusion can arise regularly and when it does it is important to spell out units clearly. Unit conversion formulae are also offered in Appendix 4.

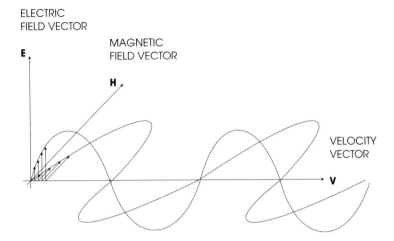

Figure 2.1   Electromagnetic wave propagation vectors.

2.2   Propagation Fundamentals

2.2.1   Electromagnetic Vectors

From basic physics, radio waves are considered to be electromagnetic in origin. Electromagnetic waves have three perpendicular vectors describing their generation. An electrical vector **E**, a magnetic vector **H** and the propagation vector **v** (Figure 2.1).

2.2.2   Polarization

A practical consideration here is the term polarity or polarization sense regarding radio wave propagation. A general convention used in radio engineering is when talking about polarization, the electrical vector is being discussed; i.e. vertical polarization is when the electrical E field is excited in a vertical direction relative to the earth's surface; horizontal polarization is when the electrical E field is excited in a horizontal direction relative to the earth's surface; and circular polarization is a hybrid of both vertical and horizontal polarity where the direction of the electrical field excitation revolves between the two planes going out along the axis of propagation (Figure 2.2).

2.2.3   Speed of Propagation and Relationship to Wavelength and Frequency

Electromagnetic waves travel at the same speed as light in a vacuum, i.e.

$$v = f\lambda \tag{2.1}$$

where v is the velocity in metres per second (m/s), f is the frequency in hertz (Hz) or cycles per second and λ is the wavelength in metres (m).

In particular in free space (vacuum conditions), v has been found to be a constant denoted by c:

$$c = 3 \times 10^8 \text{ m/s} \tag{2.2}$$

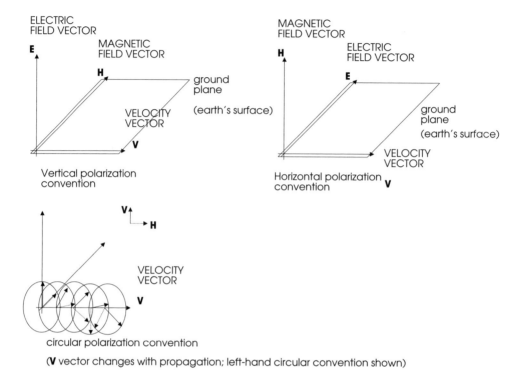

Figure 2.2   Polarization conventions.

Example 1

A high-frequency radio station broadcasts at 22.4 MHz. What is the wavelength of operation?

Answer
Using the equation

$$v = f\lambda$$

rearrange

$$\lambda = v/f$$

But electromagnetic waves propagate at the speed of light c

$$c = 3 \times 10^8 \text{ m/s}$$

So substituting c and f into the equation

$$\lambda = \frac{3 \times 10^8}{22.4 \times 10^6}$$

$\lambda = 13.39$ m (to two decimal places).

Example 2

The aeronautical high-frequency band extends from 3 to 30 MHz. Antenna design and physical lengths (which will be described later) are a function of wavelength (for example, sometimes $\lambda/4$ a quarter wave dipole, sometimes $5\lambda/8$ for some mobile applications such as GSM, etc.).

What range of wavelengths is present in the high-frequency (HF) band?
What can the physical dimensions of HF antennas be expected to be compared to those of VHF antennas? (The aeronautical communication and navigation band goes from 108 to 137 MHz.)

Answer

Part I
Apply the equation

$$v = f\lambda$$

Rearrange

$$\lambda = v/f$$

Substitute in the two values we have for frequency: for 3 MHz,

$$\lambda = \frac{3 \times 10^8}{3 \times 10^6} = 100\,\text{m}$$

and for 30 MHz

$$\lambda = \frac{3 \times 10^8}{30 \times 10^6} = 10\,\text{m}$$

So the wavelength for aeronautical HFs varies from 10 m up to 100 m.

Part II
Running the VHFs through the same equation: for 108 MHz

$$\lambda = \frac{3 \times 10^8}{108 \times 10^6} = 2.78\,\text{m}$$

and for 137 MHz

$$\lambda = \frac{3 \times 10^8}{137 \times 10^6} = 2.19\,\text{m}$$

- So in the VHF communication and navigation band of 108–137 MHz range, wavelength ranges from 2.78 to 2.19 m, respectively.
- Comparing this with the HF band, it can be seen that VHF wavelengths are typically an order of magnitude of 10 times smaller.
- Therefore it is expected that the physical dimensions of the VHF antenna systems (which will be discussed later) are a function of wavelength, which is much smaller (typically a factor of say 10 times) than their HF equivalents. This is the case.

2.3   Power, Amplitudes and the Decibel Scale

In radio, there are considerable dynamic ranges, for example, for power levels, power losses, signal voltages and field strengths. Typically, powers can be in the region of kilowatts (kW) and even megawatts (MW) for small-duty cycles of large pulsed transmitters and conversely, receiver thresholds for low-noise receivers can be down at the picowatt level. Thus the span can be seen to cover typically from 10^{-12} up to 10^6 W, to consider just one of the units.

Showing this graphically becomes very difficult on a standard scale and in addition, calculations can become cumbersome. Hence the logarithmic scale is often used: this is often called the decibel scale in radio engineering. (Note that with decibels, the logarithm to the base 10, $\log_{10}$, is used.) This leads to it often being more convenient for calculation purposes to deal in decibels for ratios or decibel units.

In general,

for *power*

$$dB(\text{unit x}) = 10\log_{10}(\text{unit x}) \tag{2.3}$$

$$\text{Power in dBW} = 10\log_{10} P(\text{watts}) \tag{2.4}$$

Sometimes a more useful value for power is dBm

$$\text{Power in dBm} = 10\log_{10} P(\text{milliwatts}) \tag{2.5}$$

$$\text{Power in dBm} = 10\,\log_{10} P(\text{watts}) + 30 \tag{2.6}$$

and for *amplitudes* (e.g. voltage, current, field strength)

$$dB(\text{unit x}) = 20\log_{10}(\text{unit x}) \tag{2.7}$$

for example the field strength E

$$E(\text{dB }\mu V/m) = 20\log_{10} E(\mu V/m) \tag{2.8}$$

Example 3

A typical aircraft radio transmitter is said to have a peak output power at the antenna port of 16 W RMS. What is this in dBW?

Answer
Applying Equation (2.4),

Power in dBW $= 10\log_{10} 16$

Power dBW $= 12.04$ dBW

Example 4

ICAO specify[4] the minimum required field strength for VHF communications at an airborne antenna to be 75 μV/m (or 20 μV/m for a ground system) at the receiver antenna. What is this in dBμV/m?

Answer

Applying Equation (2.8),

µV/m	dBµV/m
75	37.5
20	26.0

2.4 The Isotropic Power Source and Free Space Path Loss

2.4.1 Definition of Isotropic

One of the most important concepts in radio engineering is that of an isotropic power source. By definition,[3] an isotropic power source

- is from a pinpoint source of infinitesimal low volume (i.e. volume $\approx 0\,\text{m}^3$);
- radiates radio power uniformly in all directions (i.e. the envelope of propagation is an outwardly expanding sphere);
- has no loss.

Obviously this is an ideal situation that cannot exist in practice. However, this concept is the central method of referencing nearly all radio calculations and for benchmarking most radio systems and antennas (Figure 2.3).

2.4.2 Derivation of Free Space Path Loss Equation

Consider a radio link that starts at point P_{Tx} and is being received at point P_{Rx}. Assume an isotropic transmitting source P_{Tx} (watts). Now consider the power passing through a unit aperture at distance d (metres) from the isotropic source (i.e. consider this area as separate on the surface of the propagation sphere).

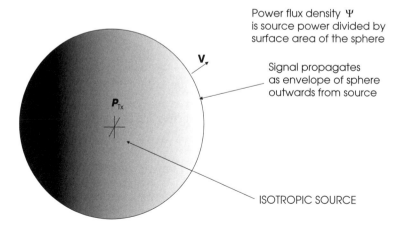

Power flux density Ψ is source power divided by surface area of the sphere

Signal propagates as envelope of sphere outwards from source

ISOTROPIC SOURCE

Figure 2.3 Radiation from an isotropic source.

The power passing through this an area on the surface of the sphere, i.e.

$$\text{the power flux density or PFD denoted as } \Psi(\text{W/m}^2) = P_{\text{Tx}}/4\pi d^2 \qquad (2.9)$$

Now consider an isotropic receiving antenna placed on this sphere. An isotropic receive antenna located on this sphere will absorb power from the radiation field it is situated in. The amount of power that the receiving antenna absorbs in relation to the RF power density of the field is determined by the effective aperture of the antenna in square metres.

For an isotropic antenna, the effective aperture A_e is given by

$$A_e = \frac{\lambda^2}{4\pi} \qquad (2.10)$$

(The absolute derivation of this equation is involved and is outside the scope of this book.)

The received power into an isotropic antenna, therefore, is

$$P_{\text{Rx}} = \frac{\lambda^2 \times \text{PFD}}{4\pi} \qquad (2.11)$$

Substituting Equations (2.9) and (2.10) into (2.11),

$$P_{\text{Rx}} = \frac{\lambda^2 \times P_{\text{Tx}}}{(4\pi d)^2}$$

Taking logs of each side of the equation,

$$10 \log_{10} P_{\text{Tx}} - 10 \log_{10} P_{\text{Rx}} = 20 \log_{10}(4\pi) + 20 \log_{10}(d) - 20 \log_{10}(\lambda) \qquad (2.12)$$

The expression $10 \log_{10} P_{\text{Tx}} - 10 \log_{10} P_{\text{Rx}}$ is the difference in what is transmitted and what is received and can be called path loss (sometimes called free space path loss); i.e.

$$L_{\text{fspl}} = 10 \log_{10} P_{\text{Tx}} - 10 \log_{10} P_{\text{Rx}} \qquad (2.13)$$

Or substituting Equations (2.12) into (2.13)

$$L_{\text{fspl}} = 20 \log_{10}(4\pi) + 20 \log_{10}(d) + 20 \log_{10}(f) - 20 \log(3 \times 10^8) \qquad (2.14)$$

A more practical form is to have distance in kilometres and frequency in MHz. This equation now becomes

$$L_{\text{fspl}} = 20 \log_{10}(4\pi) + 20 \log_{10}(d_{\text{km}}) + 60 + 20 \log_{10}(f_{\text{MHz}}) + 120 - 20 \log_{10}(3 \times 10^8)$$
$$L_{\text{fspl}} = 32.44 + 20 \log_{10}(d_{\text{km}}) + 20 \log_{10}(f_{\text{MHz}}) \qquad (2.15)$$

This is, one of the most important radio formulae. It is reproduced in ITU Recommendation PN 525-2 'Calculation of free space attenuation'.

Example 5

Part A

What is the worst-case free space path loss between an aircraft VHF transmitter and the receiving station 300 km away. Assume line of sight (LOS)!

Part B

Given a fixed transmitter power, in order to quadruple the amount of power received at the receiver, how much closer would an aircraft need to be?

Answer

Part A

Plugging the values directly into the free space path loss, Equation (2.15),

$$L_{fspl} = 32.44 + 20 \log_{10}(d_{km}) + 20 \log_{10}(f_{MHz})$$

$$L_{fspl} = 32.44 + 20 \log_{10}(300) + 20 \log_{10}(137)$$

$$L_{fspl} = 124.72\,dB$$

Part B

The new loss $L_{fspl2} = 1/4$ of 124.72 dB

$$10 \log_{10} 1/4 = -6.02\,dB$$

$$new\ loss = 118.70\,dB$$

Therefore using (2.15), the new distance

$$d = 150\,km$$

Note an interesting practical observation here: To double the distance of propagation under free space conditions, one must quadruple the power.

2.4.3 Power Flux Density

Let us consider the new concept of PFD, which can be defined as the amount of power radiated per unit area from an isotropic source at a point some distanced from the isotropic source.

From the geometry and the definition of isotropic:

$$\Psi \text{ or PFD or } (S) = P_{Tx}/4\pi d^2 \tag{2.16}$$

Power flux density can be denoted as Ψ, PFD or S.

2.4.4 Electric Field Strength

From basic physics, remember (a) Ohm's law $V = IR$ (where V is the voltage in volts, I is the current in amperes and r is the resistance in ohms (Ω)) and (b) the electrical power equation $P = IV$ (where P is power in watts, I is current in amperes and V is voltage in volts). Using the two equations above,

$$P = \frac{V^2}{R} \tag{2.17}$$

There is an equivalent analogous version for radio propagation

$$\Psi, \text{PFD}, (S) = \frac{E^2}{\text{Impedance of free space}} \tag{2.18}$$

where E is a new concept called field strength and is measured in volts per metre (V/m) and the impedance of free space (Z) is known to be a constant:

$$Z = 120\,\pi \text{ or } 377\,\Omega \tag{2.19}$$

So now a relationship between Ψ, PFD(S) and field strength E can be established. In logarithmic terms, taking $10 \times$ logs of each side:

$$10 \log S = 20 \log E - 10 \log(377).$$

$$\text{So } \Psi \text{ PFD}(S)(\text{dBW/m}^2) = E(\text{dBV/m}) - 10 \log(377).$$

Sometimes it is more useful to consider field strength in dBμV/m. Therefore the formula becomes

$$\Psi \quad \text{or} \quad S(\text{dBW/m}^2) = E\,(\text{dBμV/m}) - 60 - 60 - 25.76$$

$$\Psi \quad \text{or} \quad S(\text{dBW/m}^2) = E\,(\text{dBμV/m}) - 145.76 \tag{2.20}$$

This is also a very useful formula giving a direct relationship between field strength and PFD at the same geometrical point.

2.4.5 Relationship Between Field Strength and Transmitted Power

Continuing the theme of electrical field strength and combining Equations (2.16) and (2.18) above,

$$E^2/Z_0 = P_{\text{Tx}}/4\pi d^2$$

Taking logs,

$$20 \log E\,(\text{V/m}) = 10 \log P_{\text{Tx}} - 10 \log 4\pi - 20 \log d\,(m) + 10 \log Z_0$$

$$E\,(\text{dBμV/m}) - 60 - 60 = P_{\text{Tx}}\,(\text{dBW}) - 10.99 - 20 \log d\,(\text{km}) - 30 - 30 + 25.76$$

$$E\,(\text{dBμV/m}) = P_{\text{Tx}}\,(\text{dBW}) - 20 \log d\,(\text{km}) + 74.8 \tag{2.21}$$

Yet another important equation giving field strength from an isotropic source. (This is also covered in Section 4 of ITU-R rec PN 525-2 'Calculation of free space attenuation'.)

Example 6

- Plot a graph of electric field strength versus received power at the input to an isotopic antenna.

Use the range 0–100 μV/m. Put on the ICAO minimum field strengths already provided.

Answer

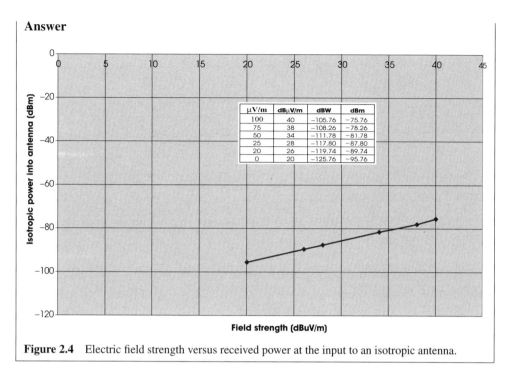

µV/m	dBµV/m	dBW	dBm
100	40	−105.76	−75.76
75	38	−108.26	−78.26
50	34	−111.78	−81.78
25	28	−117.80	−87.80
20	26	−119.74	−89.74
0	20	−125.76	−95.76

Figure 2.4 Electric field strength versus received power at the input to an isotropic antenna.

2.5 Radio Geometry

2.5.1 Radio Horizon Calculations

Consider an aircraft antenna located at a point A, at a height above mean sea level of h_1 in metres. Let the radio horizon distance from this aircraft be d_1 in metres. Let the horizon point be called point X.

The geometry of this can be drawn as below (Figure 2.5).

If O is the centre of the earth, the radius of the earth R is known to be approximately 6370 km. By definition, AXO is a right-angle triangle. So using Pythagoras theorem

$$(h_1 + R)^2 = d_1^2 + R^2$$

This can be rearranged to

$$h_1(2R + h_1) = d_1^2$$
$$d_1 = (h_1(2R + h_1))^{0.5} \tag{2.22}$$

In the expression $(2R + h)$, R is significantly larger than h and this portion of the expression can be seen to tend to or be approximated by $2R$.

Therefore

$$d_1 \approx (2Rh_1)^{0.5} \tag{2.23}$$

This formula holds true if one end of a path is approximately at sea level.

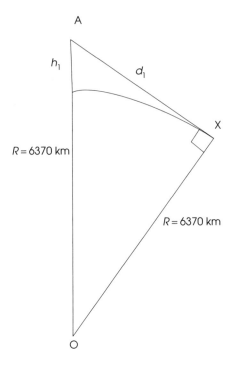

Figure 2.5   Radio horizon geometry.

Extending this argument to another airborne antenna at a point at B or a high-tower mast h_2 metres above the mean sea level, the LOS distance achievable between two points (i.e. the limit where the communication ray grazes the horizon) can be defined by

$$D \approx d_1 + d_2 = (2Rh_1)^{0.5} + (2Rh_2)^{0.5} \tag{2.24}$$

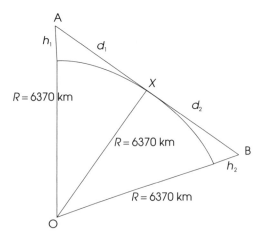

Figure 2.6   Radio horizon geometry for two aircraft or two 'high'-elevation sites.

Example 7

(a) For an aircraft flying at ceiling (assume 10 km), what is the distance to the horizon?
(b) What is the furthest apart two aircraft can be and still have LOS communications?

Answer

(a) Apply formulae

$$d = (2 \times 6370 \times 10)^{0.5}$$

$$d = 356.93 \text{ km or in aeronautical terms conservatively } 200 \text{ NM}$$

(b) For two aircraft

$$d = 713.86 \text{ or in aeronautical terms conservatively } 400 \text{ NM}$$

This example illustrates a number of interesting facts. Firstly it is no coincidence to note that in ICAO Annex 10 to the convention[4] 'Notes for guidance on VHF communications' where 200 NM is taken as the LOS horizon distance for aircraft at cruising altitude. This is based on this phenomenon described by Equation (2.23).

This 200-NM limit is also an important factor for LOS VHF communications for coverage design and for interference management considerations. Both of these are discussed later in the practical considerations section of this book.

It also explains why radio masts are placed at high sites where coverage is the principal requirement and upper airspace coverage can be extended as a function of how high the mast is. This too is a major consideration when designing coverage and is looked at in more detail in the system part of this book (Chapter 3).

Example 8

Calculate the horizon distances for each of the flight levels shown and graph the results.

Flight level	Flight level (ft)	Flight level (m)	Horizon distance (km)	Horizon distance (NM)
40				
100				
150				
200				
250				
300				
350				
400				

Answer

Flight level	Flight level (ft)	Flight level (m)	Horizon distance (km)	Horizon distance (NM)
40	4000	1219.21	124.63	67.30
100	10 000	3048.04	197.06	106.41
150	15 000	4572.06	241.35	130.33
200	20 000	6096.07	278.68	150.49
250	25 000	7620.09	311.58	168.25
300	30 000	9144.11	341.32	184.31
350	35 000	10 668.13	368.66	199.08
400	40 000	12 192.15	394.12	212.82

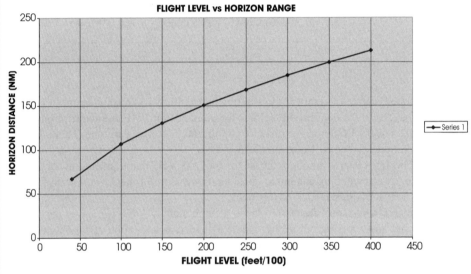

Figure 2.7   Radio horizon as a function of aircraft height. Reproduced with the kind permission of ITU.

2.5.2   Earth Bulge Factor – k Factor

In some instances, for designing radio links (particularly point–point terrestrial), it is found that radio waves do not propagate directly between two points with horizon grazing being the limiting factor as described above. They refract like light and generally tend to bend out from the earth but on occasion they bend towards the earth. This phenomenon can be described by incorporating what is called a k factor, which is used to change the apparent earth radius. This factor tends to be more significant for horizontal paths and terrestrial paths where radio rays can experience ducting (Figure 2.8).

The k factor is a multiplier for the earth's radius to give an 'effective earth's radius'. Thus for a k factor of 1, the earth's radius is considered to be unchanged. For k factors of less than 1, the earth will be considered to have a virtual radius of less than 6370 km, an exaggerated curvature for a given length of radio path, which has the same effect of considering the rays bending down towards the earth and the virtual horizon coming closer, reducing LOS propagation distance.

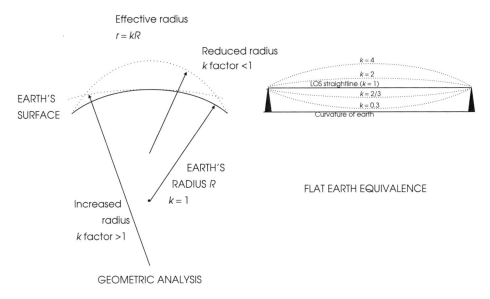

Figure 2.8 *k* factor and apparent curvature of earth.

Conversely, *k* factors of greater than 1 would simulate the earth's curvature decreasing, which is akin to rays diffracting upwards away from the earth and the horizon and subsequent LOS range of a link being pushed out.

Statistical measurements have been collected by the ITU as to the proportion of the time *k* moves up and down for different frequencies and for different climate zones, and these have been incorporated in ITU recommendations (formerly CCIR recommendations). This consideration should be incorporated in design of radio links, particularly terrestrial point-to-point links or for low-flying scenarios.

For LOS point-to-point links of high reliability, design *k* factors of 2/3 or even 0.5 are sometimes taken. By substituting into the LOS horizon equation, this can be found to reduce the horizon by typically 10–20 %.

$$\text{Apparent earth's radius} = kR \tag{2.25}$$

2.5.3 Nautical Mile

The world is split up into 360° of longitude and latitude; each degree is 60 NM and each minute is 1 NM at the earth's surface.

Definition

> Nautical mile
> A unit length used for navigation that is equivalent to 1 min of latitude extended at the earth's surface.

The nautical mile should not be confused with a statute mile as defined in the United Kingdom back in 1593 or any other mile for which there are numerous definitions, and the topic is best avoided. The nautical mile is the term frequently used by aviation. It can directly be converted to kilometres as shown in Appendix 4.

2.5.4 Great-circle Distances

The *great-circle distance* is the shortest distance between any two points on the surface of the earth (considered as approximately spherical), measured along a path on the surface of the sphere. (Height differences at each end of the path are best ignored as they are usually negligible but all the same will cause slight variation to this path.)

In fact the earth is a flattened sphere or spheroid with values for the radius of curvature of 6336 km at the equator and 6399 km at the poles. However, approximating the earth as a sphere with a radius of 6370 km results in an actual error of up to about 0.5 %.

First approximation. For a first approximation, based on the premise that the earth is a sphere, the great-circle distance is

$$D = 2R \sin^{-1}[(\sin^2((\text{lat}_1 - \text{lat}_2)/2) + \cos \text{lat}_1 \cos \text{lat}_2 \sin^2((\text{long}_1 - \text{long}_2)/2))]^{0.5} \quad (2.26)$$

The detailed derivation of this formula is quite involved mathematically and is not included in the scope of this book. It comes from straight geometry. Altitude deltas between take-off and landing heights are ignored as negligible. It should be noted that such formulae need to be applied in radians (Figure 2.9).

This formula will give an accuracy to within ±0.5 % (the error coming from the fact the earth is actually a spheroid, not a sphere) which is considered generally accurate enough for radio application. For a more accurate calculation and in some very explicit circumstances,

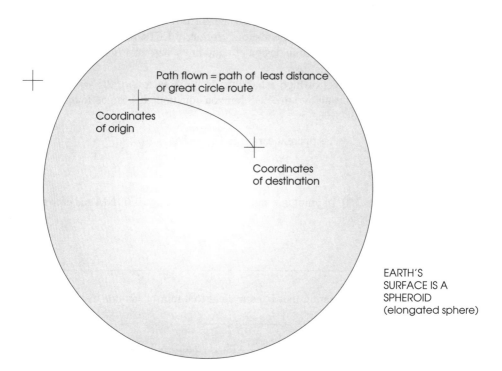

Figure 2.9 Great-circle distances.

the haversine formula is required, which allows for the non-spherical nature of the earth and corrects for these errors; however, for most purposes this is not required for radio engineering calculations.

Example 9

An aircraft reaches cruise altitude (assume 10 000 km) after departing Cairns Airport, Australia ($16°$ $36'S$, $145°$ $30'E$). At a GPS fixing of ($18°$ $18'S$, $139°$ $12'E$) the VHF communication is finally lost. The altitude of the top of the VHF mast in Cairns is 890 m. (Use a nautical mile/km conversion of 8/5.)

What does this say about the k factor?

Answer

Firstly, obtain the greater circle distance between the two points along the earth's surface by applying Equation (2.26). Now converting into radians:

$lat_1 = 0.28972;$ $long_1 = 2.53945$

$lat_2 = 0.3194;$ $long_2 = 2.4295$

$$D = 2 \times 6370 \sin^{-1}[(\sin^2((0.02968)/2) + \cos 0.28972 \cos 0.3194 \sin^2((0.10995)/2))]^{0.5}$$

$$D = 694.32 \text{ km}$$

On initial inspection this looks way beyond the reasonable radio horizon, so some kind of bending of the radio beam upwards (akin to k factors in excess of 1) must be taking place.

By broad definition this is the point where coverage ceases, so it should be assumed this is where LOS stops.

Consider the k factor. Using Equations (2.24) and (2.25)

$$D = d_1 + d_2 = (2kRh_1)^{0.5} + (2kRh_2)^{0.5}$$

Solving for k

$$k = 2.8$$

This is seriously adverse propagation in favour of increasing range.

It should be noted that with VHF, frequently signals can be heard from transmitters that are well over the horizon, strengthening the theory of the k factor. (This is a common problem experienced in most parts of the world; it is discussed in more detail in the practicalities section of the book, Chapter 12.)

Some statistical analysis has been carried out to look at the variance of k with time. More detailed study is contained in the ITU-R recommendations, P (for propagation) archives.

2.6 Complex Propagation: Refraction, Absorption, Non-LOS Propagation

Discussion has so far been mainly concentrated on how radio waves propagate in free space with LOS conditions. This describes the majority case for aviation and most situations can be approximated to this.

Occasionally, however, an aircraft will not have LOS to the nearest ground node. This is usually the case on a long approach, or approach in mountainous terrain or when the station is

REFRACTIVE LAYERS
(SOMETIMES CALLED DUCTS)
(each layer has a different gas density and hence refractivity index)

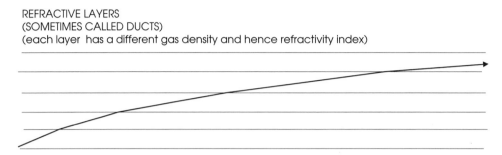

Figure 2.10 Incremental refraction.

over the horizon. At such lower altitudes other propagation mechanisms can come into effect, such as

- reflection;
- refraction;
- ground wave/sky wave.

2.6.1 Refraction

2.6.1.1 Layer Refraction

The k factor starts to introduce the concept of non-straight-line radio beams. This is essentially incremental refraction occurring at every point along the beam. Much study and prediction of this has been carried out by ITU-R and archived in the P (propagation) archives (in particular 'ITU-R P-453-9[5] – The radio refractive index: its formula and refractivity data') (Figure 2.10).

Of course, in the limit, the phenomenon of total internal refraction or 'ducting' occurs and a radio beam can extend well beyond its natural horizon limits (Figure 2.11).

REFRACTIVE LAYERS
(SOMETIMES CALLED DUCTS)
(each layer has a different gas density and hence refractivity index)

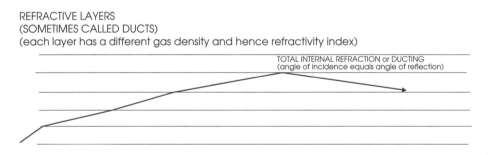

Figure 2.11 Total internal refraction or 'ducting'.

2.6.1.2 Obstacle Refraction

A second type of refraction that occurs is that when the radio wave (showing its wave physics properties) is blocked by either a smooth obstacle that encroaches into its path or a knife-edge obstacle encroaching into its path. An attenuation of the beam occurs, which is a function of the Fresnel physics (Figures 2.12–2.14).

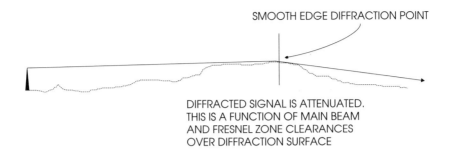

SMOOTH EDGE DIFFRACTION POINT

DIFFRACTED SIGNAL IS ATTENUATED.
THIS IS A FUNCTION OF MAIN BEAM
AND FRESNEL ZONE CLEARANCES
OVER DIFFRACTION SURFACE

Figure 2.12   Smooth obstacle refraction.

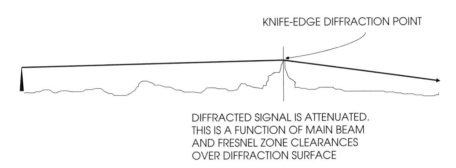

KNIFE-EDGE DIFFRACTION POINT

DIFFRACTED SIGNAL IS ATTENUATED.
THIS IS A FUNCTION OF MAIN BEAM
AND FRESNEL ZONE CLEARANCES
OVER DIFFRACTION SURFACE

Figure 2.13   Knife-edge diffraction.

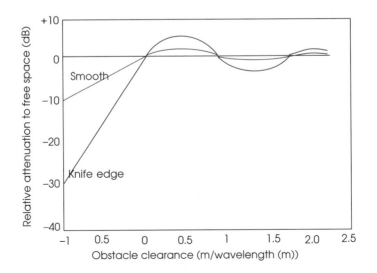

Figure 2.14   Attenuation as a function of geometry.

2.6.2   Attenuation from Atmosphere Absorption

Consider the molecular properties of radio waves and the concept that radio waves travel through gaseous medium. As the radio signal is promulgated through this medium, the gases take some of the energy out of the radio wave, particularly if the resonant frequency of the gas molecule under consideration is at or near the frequency of the radio wave. This phenomenon is called *absorption*.

2.6.2.1   Water Absorption

For terrestrial paths operating below about 5-GHz frequency multipath, flat fading or selective fading (sometimes called fast fading) is the significant contributor to degradation from the free space path loss condition (or fading as it is called). The gaseous absorption loss is usually negligible. Above the frequency of 5 GHz, water absorption particularly becomes more significant and usually is the dominant mechanism for fading on a terrestrial path.

The proportion of the time for which the severe attenuation occurs is usually only when the path is subject to medium to heavy rain. The topographic rain profile of an area is an important feature to such paths. ITU have studied these rain profiles and categorized each area of the world with a profile letter designation. (CCIR have rain zone tables and prediction methods; see rain zones described in ITU-CCIR Report 563-2.)

In particular the temperate climates are found to be less prone to rain fading than are the tropics, e.g. Northern Europe ranges from zone C to G, the subtropics are ranging typically to K and the tropics can go from M up to W and X.

When designing links or paths, consideration of the rain zones should be given. For ICAO the worst rain zone attenuations usually have to be considered when designing radio systems for operation in all regions of the world (Figure 2.15).

2.6.2.2   Oxygen Absorption and Other Gases

Of lesser significance is absorption from other atmospheric gases, particularly oxygen. The frequencies of 6 GHz and 13 GHz are where oxygen absorption peaks (Figure 2.16). (See CCIR Report 719.)

At lower HF and VHF frequencies gaseous absorption is negligible.

CCIR rain region (Previous designation)	Example area	Worst hour (mm/hr)
A	Arctic/Antarctic	<10
B	Greenland	10
D	UK South East	25
E	UK rest	30
K	Sydney	40
M	Brisbane	60
–	Rio de Janeiro	70.4
W	Cairns	80
–	Cameroon	100
–	Singapore	120

Figure 2.15   Rain attenuation curves. For fuller description of how to apply formulaes and principles and attain dB/km, see ITU-R P837.

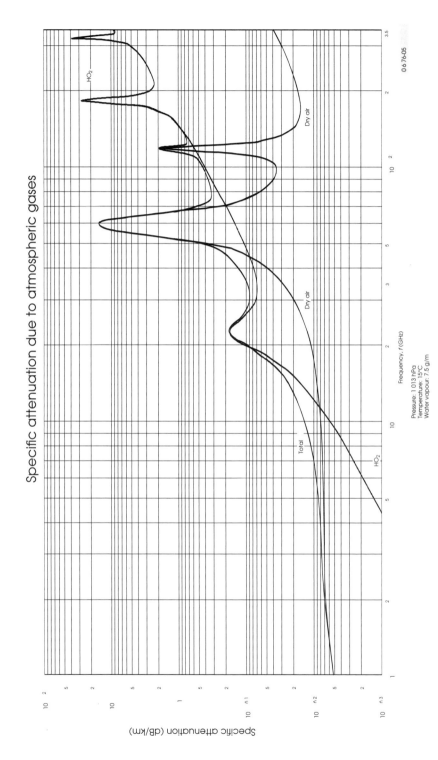

Figure 2.16 Gas absorption. Reproduced with the kind permission of ITU.

2.6.3 Non-LOS Propagation

There are a number of other mechanisms for which propagation can happen in non-direct LOS. These are via ground wave propagation (which tends to be more significant at low or very low frequencies, in kilohertz or less), reflection and refraction (which are both related and similar), which is prevalent in the HF, VHF and UHF bands. In the following, these will be considered in more detail.

2.6.3.1 Propagation – Ground Wave

This is when radio waves follow the curvature of the earth. This is particularly the case for low frequencies and very low frequencies in the kilohertz region. Ground waves that follow the earth must be vertically polarized. The attenuation is a function of frequency and is lower at lower frequency.

Generally ground wave propagation is not appropriate for aeronautical purposes, either in frequency (the aeronautical bands are too high in frequency) or application (Figure 2.17).

2.6.3.2 Reflection and Multipath

HF, VHF, UHF and microwave frequencies lend themselves easily to reflection, refraction and multipath. This is when by the law of physics, a ray is incident to a 'reflective surface' (and this can be, for example, still water, land, metal or a highly conductive surface). In undergoing a reflection, usually some losses are incurred but these can be small, in which case a reflected ray can be comparable with a direct ray in terms of incident power (Figure 2.19). From a practical perspective, this is the reason why these frequencies are often chosen for radar systems, which exploit this property.

For mobile communications, in some cases, this can be a favourable property, extending range and saturating coverage for certain systems (i.e. cellular mobiles) in a cluttered urban environment where frequencies of 400 and 800 MHz are well known for their good properties of bouncing off walls and propagating through windows and tunnels.

Also for aviation the property can be useful, particularly at HF where relatively low-loss reflections can be used to provide long-range radio. However, in other aeronautical applications, particularly between low-altitude transceivers, usually it can also lead to severe attenuations even when direct LOS exists and is a very undesirable phenomenon.

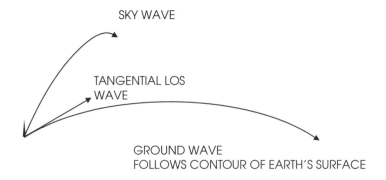

Figure 2.17 Ground wave propagation.

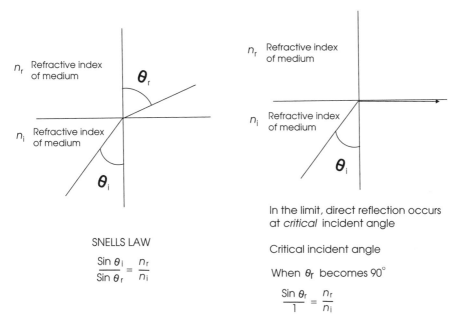

Figure 2.18 Direct refraction.

There are many models derived for terrestrial non-LOS, usually based on the two-ray model or the extended and advanced work on this theme by Hata and Okumura.

2.6.3.2.1 The Two-Ray Model

Consider an instance where two signals (one usually direct and one indirect) from the same source arrive at a receiver. Constructive or destructive fading can take place, resulting in either good or bad signal reception (Figure 2.19).

This phenomenon is called multipath and becomes very significant for marginal LOS cases or radio communications close to the earth's surface, particularly over water or in metallic cluttered environments such as around airports. Of course the theme can be extended where multiple indirect rays can be arriving at the receiver with or without a direct ray, compounding the complexity and uncertainty of the situation.

Hata and Okamura extensively studied this phenomenon regarding its applications to mobile communications. It can even be exploited thoroughly in some instances, but generally it is a nuisance and countermeasures are used to cancel the destructive interference aspects (using

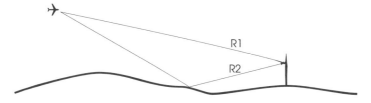

Figure 2.19 Reflection and the two-ray model. (Direct (R1) and indirect (R2) ray.)

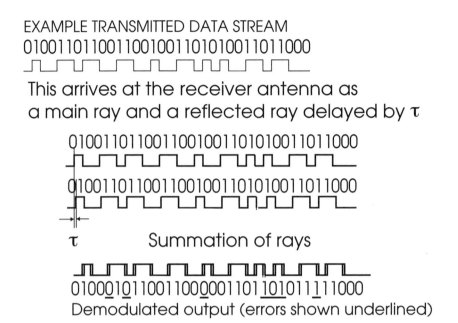

EXAMPLE TRANSMITTED DATA STREAM
010011011001100100110101001101100

This arrives at the receiver antenna as
a main ray and a reflected ray delayed by τ

010011011001100100110101001101100

010011011001100100110101001101100

τ   Summation of rays

01000101100110000011011010111110000
Demodulated output (errors shown underlined)

Figure 2.20   Intersymbol interference.

equalization techniques, delay equalization processing or data interleaving techniques, to name just a few of the system measures to be discussed later.)

When multipath occurs and the delay period is less than a data period between two symbol states in the digital environment, intersymbol interference will occur. This can be seen when observing a data stream at the intermediate frequency (IF) level through an 'eye diagram'. The eye will close as a function of how much reflected signal is arriving at the receiver and the intersymbol interference builds. Figure 2.20 describes example transmitted and received data stream in a multipath environment. Figure 2.21 shows how an eye diagram can be obtained.

2.6.3.3   Propagation – Sky Wave

Sky wave propagation is an extension of the reflection theme. The waves that rely on the refractive properties of the ionosphere to propagate long distances are considered as sky waves (Figure 2.22). (In the limit, a refraction becomes a reflection; and hence the relationship to reflections.) This phenomenon is very frequent below 30 MHz. These properties are useful in that long-range communications can be provided well over the horizon. The HF band (3–30 MHz) is well suited for this and aviation and other radio users exploit this band for this very purpose. Conversely the reflective properties of the ionosphere can occur into the low VHF band (above 30 MHz) and even beyond for a low proportion of time into the aeronautical communications band (currently 118–137 MHz). This is usually more of a problem where coverage can over-reach past the horizon into an adjacent area sometimes up to 1000 km away using the same frequency for a different application of airspace and consequently causing interference.

HF traditionally is considered to extend from 3 to 30 MHz. The aeronautical HF band, however, is considered to extend a little before this to start at 2 MHz and run up to 22 MHz.

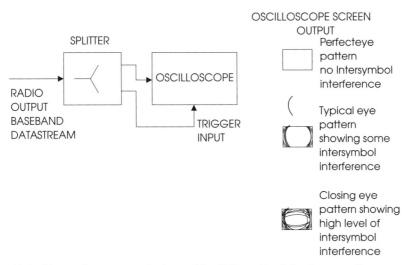

Note it is usually necessary to turn off the FEC and/or Adaptive equaliser in the receiver to properly see the 'eye' pattern

Figure 2.21 Obtaining the 'eye diagram'.

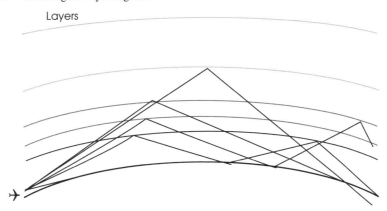

Figure 2.22 Sky wave propagation.

In the early days of radio, HF was a favourite band because the propagation properties are such that under certain conditions the wave can totally internally reflect or 'skip' its way right around the earth. In addition, the antenna properties are very efficient and finally, the transmitter/receiver was easily designed and economic. Today, this still holds true.

2.6.3.3.1 The Ionosphere

The ionosphere extends nominally from 60 to 400 km above the earth's surface. It is so called as the particles in this region are easily ionized, giving it special electrical properties.

Under certain conditions the HF radio waves will refract and in the limit totally internally reflect or 'duct' (Figure 2.23).

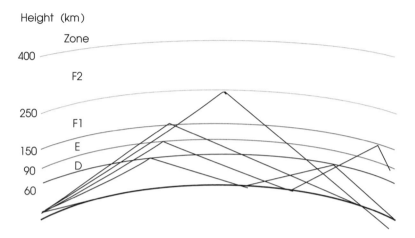

Figure 2.23 HF reflection from the different layers.

Theoretically, there is a number of layers that exist. The D layer runs from height 60 to 90 km and the E layer from 90 to 150 km. The F1 layer runs from 150 km to 250 km and the F2 layer from 250 km to 400 km.

Daytime. In general, the lower layers absorb lower frequencies typically up to 10 MHz during the daytime, and the higher frequencies typically 10–30 MHz are reflected off the F1 layer (Figure 2.24).

Night Time. In general, the D and E layers disappear and the F layer reflects frequencies below typically 10 MHz. (The higher frequencies pass straight through the ionosphere without refracting/reflecting).

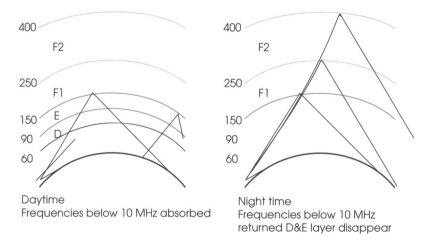

Figure 2.24 Daytime and night time operation.

Unfortunately, with HF there is an element of 'luck' or randomness to these general rules. ITU recommendation P.533-7 –'HF Propagation Prediction Method', gives an interesting insight into this. It provides a theoretical method to calculate which frequencies are most suitable at which times. It identifies a couple of concepts.

Maximum Usable Frequency. This is a median value statistically. The chances at the time of validity of an HF channel frequency being 'open' above this value are 50 %.

Highest Possible Frequency. This is the HF, above which statistically there is only a 10 % chance of the frequency being open.

Optimum Working Frequency. This is the highest HF that gives a 85 % likelihood of the channel being open.

Critical Frequency (FOE). This is the highest frequency that is returned in ionosphere. This can be applied for all path lengths from hundreds of kilometres up to beyond 9000 km. (This is effectively half-way around the world, which is as far as you can practically go.) This theory also investigates the various modes discussed above in detail (e.g. E modes up to 4000-km range and F2 modes for all distances).

There is a complex formula given in this recommendation

$$L_t = 32.45 + 20 \log f + 20 \log p\varphi - G_t + L_i + L_m + L_g + L_h + L_z \qquad (2.27)$$

where

$$p' = 2R_0 \sum_1^n \left[\frac{\sin (d/2R_0)}{\cos [\Delta + (d/2R_0)]} \right]$$

L_i is the absorption loss (dB) for an n-hop mode, given by

$$L_i = \frac{n(1 + 0.0067 R_{12}) \sec i}{(f + f_L)^2} \frac{1}{k} \sum_{j=1}^k AT_{noon} \frac{F(\chi_j)}{F(\chi noon)} \varphi \left(\frac{f_v}{foE} \right)$$

Note the resemblance to the free space path loss equation; also, there is a time delay element.

Time delay – The time delay of an individual mode is given by

$$\tau = (p'/c) \times 10^3 \text{ ms}$$

where p' is virtual slant range (km) and c is the velocity of light (3×10^8 m/s).

Practical Measurements – Ionosphere Sounding. This is a technique frequently used in practice to gauge the current ionosphere conditions and to be able to predict propagation. In its simplest form, consider sending a multifrequency signal straight up perpendicularly to earth's surface and measure which frequency components are returned.

A number of services send out 'squitters' or multifrequency components like this in their transmissions, and receivers can look at what they receive and analyse, and from this evaluate what the current propagation conditions are and which frequencies are best to use at that moment in time. In reality, this practical approach is found to be a more powerful method than the theory.

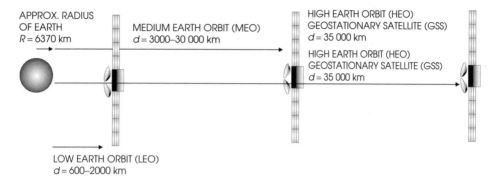

Figure 2.25   Satellite propagation.

Solar Flares and Sun Spot Activity.   Solar flares present total ionization of a specific frequency and thus can wipe out HF communication ability on a particular channel for minutes or hours. There is a random nature to their occurrences, although sun spot activity can be correlated with 11 year solar cycles.

2.6.4 *Propagation to Satellite* (Figure 2.25)

2.6.4.1 *Propagation Distance*

Probably the most significant factor for satellite communications is the distance between the satellite and the earth (or the distance between satellites for intersatellite communication). This introduces significant delay time between transmitted and received signals. Also this has a direct bearing on the path attenuation (from the free space path equation derived earlier, see Equation (2.15)), which tends to be relatively large in satellite communications.

2.6.4.2 *Atmospheric Losses*

By contrast, satellite beams cut up through the atmosphere and quickly pass through the absorption/precipitation zone (usually this is less than 1 % of their path length) (the exception being at the poles to geostationary satellites (above the equator) where precipitation is much less anyway) and so rain and gaseous attenuation generally becomes insignificant.

For aeronautical applications, the main frequency bands of interest to aviation are L band 1.4 GHz C band 4 and 6 GHz and Ku band 12–14 GHz, where the losses are minimal for satellite paths and international allocations have been made.

Doppler effects can be significant to the radio systems at the satellite end, as a satellite can be manoeuvring at 'galactic' speeds to maintain its orbital position.

There are generally three types of satellites:

High earth orbit (HEO) or sometimes called *geostationary earth orbit* satellites (GSS) are the ones that are in stationary (relative to earth movement) orbit at nominally 35 000 km above the equator. That is where the gravitational pull or centripetal force on the object balances or equals the centrifugal force of the object spinning with a period of one day. By the nature of physics, there is a time delay of 1/8 of a second in a radio beam travelling from the earth's surface to such a satellite. These satellites are used extensively for a number of communications applications by civil aviation, military and for entertainment broadcast and passenger services and these will be discussed in Chapter 5 in more detail. These are the most common satellites presently used by civil aviation.

Two other kinds of satellites are *medium earth orbit* (MEO) and *low earth orbit* (LEO) satellites. These fly closer to the earth and consequently are moving relative to the earth's surface and will be perceived to go though a trajectory from one horizon to the other when being observed on the ground (sometimes within a few minutes). The lower the satellite the shorter the trajectory period.

Also of consequence is that the latency or signal delay to these satellites is much reduced and the frequency reuse is much higher due to the ability of multiple satellites to be shielded from each other by the earth itself. These satellites are not currently used by aeronautical applications (other than by exception the military ones); however, they could be used in future applications where they are more attractive for latency/voice critical applications.

2.7 Other Propagation Effects

2.7.1 The Doppler Effect

In its simplest form the Doppler effect is when an object with vector velocity relative to a receiver is transmitting a certain frequency that is perceived to be different due to either the waveforms being compressed (if the vector is towards the receiver) or elongated (if the vector is away from the receiver).

Remember the school physics example of a fire engine coming towards a person with its siren on and as it passes and moves away, the note or tone apparently drops a pitch. The same applies to radio systems.

Consider Equations (2.1) and (2.2)

$$v = f\lambda \tag{2.28}$$

$$c = 3 \times 10^8 \text{ m/s} \tag{2.29}$$

Consider a stationary receiver receiving waves from a stationary source. Within a given second the number of cycles received at the receiver would be received cycles $= c/\lambda$.

Now consider a source moving towards the receiver, and received cycles $= (c + y)/\lambda$ where y is the speed in metres per second of source.

$$\text{Therefore } f_{apparent} = f_{carrier} \times (c + y)/c \tag{2.30}$$

This formula is valid for transmitting sources travelling towards or way from the target. The argument can be further extended if both the transmitter and receiver are moving.

2.7.1.1 Example

Consider the likely changes due to Doppler of carriers for various aeronautical systems. In particular, consider Table 2.1. Complete the matrix calculating the apparent changes of frequency possible for each of the aircraft with actual frequencies and (% of carrier in brackets), scenario for each of the bands being considered. Comment on what should be incorporated into the receiver design (Tables 2.1 and 2.2).

Table 2.1 Doppler shift (question).

Aeronautical system	Nominal carrier frequency	Air/ground maximum shift for jet at 1000 km/h	Air/air maximum shift for jets at 1000 km/h	Air/air maximum shift for Concorde at 2500 km/h	Satellite shift, assume maximum speed 10 000 km/h
HF	3 MHz				NA
HF	30 MHz				NA
VHF	117.975 MHz				NA
VHF	137 MHz				NA
L Sat band	1.4 GHz	NA	NA	NA	
Ku Sat band	14.5 GHz	NA	NA	NA	

Table 2.2 Doppler shift (answer).

Aeronautical system	Nominal carrier frequency	Air/ground maximum shift for jet at 1000 km/h	Air/air maximum shift for jets at 1000 km/h	Air/air maximum shift for Concorde at 2500 km/h	Satellite shift, assume maximum speed 10 000 km/h
HF	3 MHz	3 000 002 778 2 999 997 222	3 000 005 556 2 999 994 444	3 000 013 889 2 999 986 111	NA NA
HF	30 MHz	3 000 002 778 2 999 997 222	3 000 005 556 2 999 994 444	3 000 013 889 2 999 986 111	NA NA
VHF	117.975 MHz	1 180 001 093 1 179 998 907	1 180 002 185 1 179 997 815	1 180 005 463 1 179 994 537	NA NA
VHF	137 MHz	1 270 001 176 1 269 998 824	1 270 002 352 1 269 997 648	127 000 588 126 999 412	NA NA
L Sat band	1.4 GHz	NA NA	NA NA	NA NA	1 400 012 963 1 399 987 037
Ku Sat band	14.5 GHz	NA NA	NA NA	NA NA	1 450 013 426 1 449 986 574

2.7.1.2 Answer

As can be seen from the results (even with nine decimal places taken) the Doppler effect has generally negligible effect on frequency accuracy. This is well catered for within the receive channel definition as shall be shown later.

2.8 Modulation

Modulation is when the signal message that is to be conveyed is transposed (e.g. voice or raw computer data) into a suitable form so that it can be transmitted over the media involved (usually plain air with absorbing gases as described earlier).

The raw data or voice signal (usually called the baseband) can be modulated onto a 'carrier frequency' (Figure 2.26).

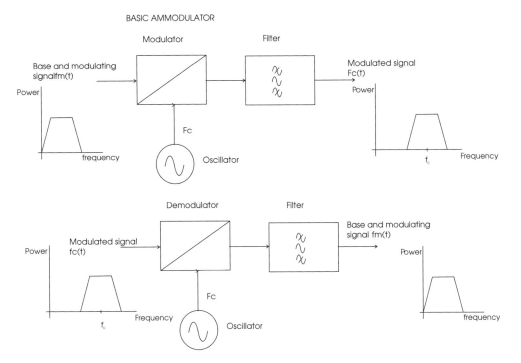

Figure 2.26   Modulation and demodulation.

A carrier frequency is selected to suit propagation conditions, spectrum and frequency management constraints and ITU regulations and efficient and reliable transportation of the information across the media. There are mathematical equations that describe this modulation process.

The reverse process 'demodulation' is when the useful information is extracted from the unuseful carrier.

There are many different types of modulation schemes available to the radio designer. In principle, these can be broken down into analogue modulation such as amplitude or frequency modulation (AM or FM) schemes (or variants on these), as can be used to modulate voice signals. Or more recently, digital modulation such as frequency shift keying (FSK), phase shift keying (PSK) or advanced variants of these can be used to convey digital messages and/or digitized voice streams, or of course both.

The selection of appropriate modulation schemes for an application are usually dependent on (but not necessarily limited to) the following factors:

1. required range of propagation;
2. frequency of operation and propagation properties;
3. spectral efficiency;
4. equipment complexity, reliability, size weight and cost;
5. required data or voice throughput rate;
6. regulatory constraints.

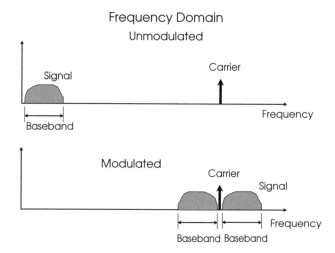

Figure 2.27 The modulation conundrum.

2.8.1 *The Modulation Conundrum* (Figure 2.27)

There is a relationship between all of these aspects (mathematically it is Shannon's law; see later), and it is usually necessary to trade off one requirement against the others to achieve the best overall engineering solution.

In reality, for low-specification systems where performance is not so critical, rugged and rudimentary low-order modulation schemes can work very efficiently, and this keeps cost and complexity down. For some of the more sophisticated systems where data throughput in a limited bandwidth is premium, some of the more sophisticated and elaborate modulation schemes pushing the limits of Shannon's law become applicable if not essential.

2.8.2 *The Analogue and Digital Domains*

Before going into the details of the modulation schemes employed by aeronautical systems, it is worth momentarily looking at the analogue to digital (A/D) aspects of signal modulation.

Modern communication systems generally digitize voice and images and transmit them via a digital medium (where quality can generally be kept) before putting them back in their original analogue domains. With data there is never any A/D and it stays entirely in the digital domain. However, the modulation process can be considered as the inverse of this process; i.e. data is modulated onto a carrier conversion (which is akin to D/A conversion) and then it is transmitted over the media and converted back to digital via A/D conversion at the end (Table 2.3).

Table 2.3 Analogue and digital domains.

Analogue to analogue conversion (anologue speech to analogue radio)	Digital to analogue conversion (data or digitized speech to radio)
Amplitude modulation Frequency modulation	Amplitude shift keying Frequency shift keying Phase shift keying

Table 2.4 Analogue and digital modulation.

Analogue	Digital
DSB-AM	PSK
SSB-AM	FSK
SC-AM	DPSK
FM	DFSK
	QAM
	TCM

With voice over legacy radio systems, it is never even coming into the digital domain. It is analogue when it starts, it goes through an analogue to analogue conversion as it is modulated and then at the end it is an analogue to analogue demodulation process.

2.8.3 Amplitude Modulation (AM)

AM is when the amplitude of the carrier is directly proportional to the modulating signal. Probably the simplest form of AM is double side band amplitude modulation (DSB-AM).

2.8.3.1   DSB-AM (Figure 2.28)

DSB-AM is described mathematically as

$$f_c(t) = A\cos(2\pi f_c t + \varphi)\pi$$

where $f_c(t)$ is the modulated signal output.

Let

$$A = K + f_m(t)$$

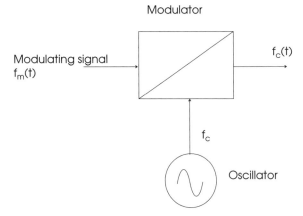

BASIC AM MODULATOR

Modulator

Modulating signal
$f_m(t)$

$f_c(t)$

f_c

Oscillator

Figure 2.28   DSB-AM modulation.

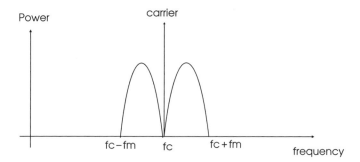

Figure 2.29   Power spectrum for DSB-AM.

where K is the unmodulated carrier amplitude and $f_m(t)$ is the baseband signal and

$$f_m(t) = a\cos(2\pi f_m t)$$

So $f_c(t) = (K + a\,\cos(2\pi f_m t))\cos(2\pi f_c t + \phi)$

$$m \text{ (depth of modulation)} = \frac{\text{Modulating signal amplitude}}{\text{Unmodulated carrier amplitude}} = \frac{a}{K}$$

Now, let $\Phi = 0$. Then multiplying out the cos function,

$$f_c(t) = K[\cos(2\pi f_c t) + 0.5\,m\,\cos(2\pi(f_c - f_m)t) + 0.5\,m\,\cos(2\pi(f_c + f_m)t)]$$

In the frequency domain, these discrete components can be seen and are called upper and lower sidebands, respectively (Figure 2.29).

2.8.3.1.1   Observations of DSB-AM

It is interesting to note that the modulated bandwidth of DSB-AM is twice the size of the minimum baseband bandwidth or the frequency component. It is independent of depth of modulation.

2.8.3.1.2   Advantages of DSB-AM

Following are the Advantages of DSB-AM:

- It is very easy to synthesize DSB-AM (i.e. non-linear device such as diode or transistor) (Figure 2.30).
- It is very easy to demodulate (non-coherent detection uses a simple diode, capacitor, resistor circuit or can use a coherent detector) (Figure 2.31).
- It is incredibly simple to work with, troubleshoot and see the waveforms at every stage of the process.
- There is always a carrier present, even when no information is being sent; this means a receiver can easily tune to this carrier frequency and keep locked to it.
- In the amplitude domain, signals cannot get distorted or shifted by Doppler shifts of receivers and transmitters that are physically moving with relation to one another.

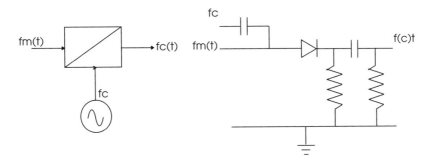

Figure 2.30 DSB-AM modulator.

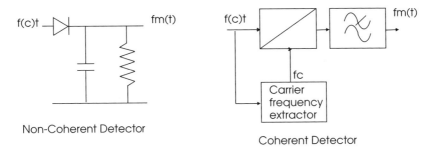

Figure 2.31 DSB-AM non-coherent detector and coherent detector.

2.8.3.1.3 Disadvantages of DSB-AM

Some of the relative disadvantages of DSB-AM when comparing this modulation to other schemes can be the following:

- At its best performance, only one-third of power being transmitted is the information signal, the other two-thirds of power is absorbed by the carrier and is not useful. This is with a modulation index of $M = 100\%$, which in theory can never be fully reached, and so the practical realization of this is even worse, with typically always less than 25% power transfer efficiency.
- DSB-AM uses more RF bandwidth than is actually needed. As was previously pointed out, the RF bandwidth required is double the maximum baseband frequency component. More elaborate modulation schemes can half this or reduce it even more.
- There is no encryption of the signal, so it is easy to interfere with (security, etc.).

2.8.3.2 *The VHF Aeronautical Mobile Communications (Route) Service (AM(R)S)*

The AM(R)S uses DSB-AM, mainly for legacy, robust and simplicity reasons. It was implemented when the spectrum resource was not a premium. Its basic building blocks are shown in Figure 2.32.

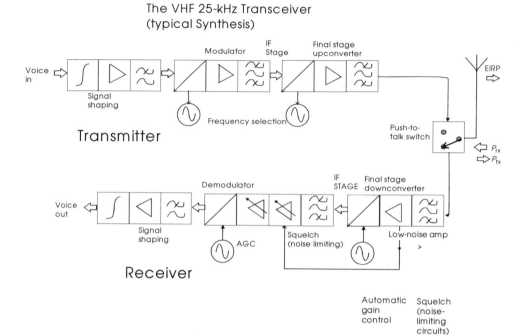

Figure 2.32   The VHF AM(R)S transceiver.

2.8.3.2.1   The VHF 25-kHz Transceiver

Figure 2.32 gives the system block layout of a typical receiver. 'Typical' is emphasized as increasingly much of this traditional block approach can be synthesized directly with digital large-scale integration techniques, i.e. all on one printed circuit, and the symbols used in the block diagram are typical of electronic engineering block diagrams.

Transmitter Path.

Signal Shaping.   Starting at the mouthpiece, the voice as spoken by pilot/controller/other is converted into electrical speech signals. These are compressed. 'Companding' (compressor/expanding) is a standard technique for most voice communications. It standardizes the dynamic range experienced in volume, pre-emphasizing quiet speakers and attenuating loud speaking people. Signals are compressed on entry into a communication system and expanded back, in theory, to where they were on the way in. This technique also offers improvement in overall noise reduction and consequential SNR improvement.

Figure 2.33   RC circuit.

The next stage is a voice amplifier. The signals are then passed into a low-pass filter. The low-pass filter which usually limits the voice signal from a few hertz (the lowest the normal voice goes to is around 80 Hz) up to 4800 Hz. (The highest pitch a normal voice goes up to is a few tones in the 10–15-kHz region; however, removing these high-pitch frequency components has minimal impact on the intelligibility of the voice, and by the casual user no appreciable difference can be discerned.) By limiting the upper frequency, ultimately the RF bandwidth is curtailed. The filter used here is usually a raised cosine filter, which has a relatively slow cut-off frequency above 4800 kHz.

Modulator. The refined voice signal (baseband signal) is mixed with the (intermediate) carrier frequency of the radio. This frequency is synthesized typically with a crystal oscillator that provides highly stable frequency sources. The synthesis circuitry is controlled by the frequency selector interface on the radio. This sets the integer frequency of operation required.

The combining of the carrier and baseband is usually done in a non-linear device such as a diode or, more commonly, in an active transistor circuit. The resultant signal is amplified and the unwanted modulation products – sometimes called distortions – are removed by the bandpass filter. The result is an IF stage.

IFs are sometimes, but not necessarily always, used as the first part in a two-part modulation process to facilitate easy filtering out of unwanted products and a good quality of modulation.

Final-Stage Upconverter. The IF signal is now mixed with the carrier frequency, using the formula

$$f_c = 117.975 + (n \times 0.025) - f_{IF} \text{ MHz},$$

to get the RF exact channel signal.

Again this signal is amplified by an RF power amplifier with typical powers around 25 W (RMS) and cleaned up by the bandpass filter, limiting what is going to air to be in the 25-kHz RF channel only. Again a raised cosine filter is used. This clean signal is then sent in the case of an aircraft dual-simplex system to the push-to-talk switch and antenna system. (In the more advanced signal processing transceivers, this stage is sometimes bypassed and baseband is modulated direct to RF carrier.)

Push-to-Talk Switch. The aeronautical broadband very high frequency system operates in simplex or half-duplex mode; i.e. one RF channel is time-shared between radio transceivers. When a pilot/controller/other user wants to talk, they press the push-to-talk button, which provides a direct path between the power amplifier stage of the upconverter and the antenna system, and the receiver is momentarily cut off from receiving signals at the same time.

In contrast when the push-to-talk switch is not pressed, the transceiver automatically connects the antenna system to the receive branch and the transmit path is fed into a dummy load (for simplicity, this is not shown in block diagram). Thus the antenna system can only be connected to one or the other (transmitter or receiver) at any given instance.

For ground equipment sometimes the two radio parts are two seperate systems.

Receiver Path. In simple terms, this is the reverse of the transmitting operation.

Final-Stage Downconverter. When the push-to-talk switch is left in idle (non-pressed) mode, the incoming RF signal travels from the antenna through the coaxial cables and connectors to the bandpass filter.

The bandpass filter excludes all the other RFs coming down the antenna except for that in the nominal VHF range (approximately 118–137 MHz). This 'cleaned' signal is then amplified and passed to the first demodulator stage.

In the first stage, the incoming signal is mixed with the carrier frequency less the IF $f_c - f_{IF}$ to 'downconvert' to a resultant IF signal.

This provides the selectivity to capture the specific RF channel wanted.

Demodulator Stage. This signal is then cleaned again by a bandpass filter to remove any unwanted products from non-linearities in the mixing and passed to a noise-limiting 'squelch' circuit.

The function of the squelch circuit is simply to act as a receiver on and off switch; i.e. if there is sufficient signal coming through the downconverter circuit, it will 'switch on' the receiving amplifier so that sound can be heard. If the signal coming down the downconverter is below a certain threshold and is just background noise or distant aircraft operating over the same channel over the horizon, the amplifier remains switched off. This is to make listening more comfortable for the pilot/controller/other user; i.e. they do not need to listen to background noise and static, but only their signals when they come through.

Next, the signal is demodulated. Demodulation is the reverse of modulation and can be performed coherently or non-coherently (the simplest form) with a simple RC circuit (Figure 3.33).

Signal Shaping. The signal is passed through a low-pass filter to take away or de-emphasize some of the higher frequency crackles and noises. It is then amplified and 'expanded' – which is the reverse process of compressing (discussed above) – and passed to a loudspeaker or headset as applicable.

2.8.3.3 Single Sideband (SSB) Modulation

With DSB-AM, it is found that the upper sideband is a perfect reflection of the lower sideband; i.e. the same information is carried by both. By removing one of the sidebands and by removing the carrier, therefore, no information is lost and the SNR (signal-to-noise ratio) increases for a given transmit power as all the transmitted energy is concentrated into the information signal. This is the basis for single sideband amplitude modulation (SSB-AM) (Figure 2.34).

The construction of an SSB is fractionally more complicated than that of DSB-AM. It can be in synthesized two different ways.

Method 1. As shown in Figure 2.35, a SSB signal can be synthesized either by using basic DSB-AM or DSB-SC-AM modulator and filtering out the unwanted carrier and sideband (usually done at a IF).

Method 2. Arguably a more elegant way is to use Hilbert modulator. This essentially cancels out the unwanted sideband and carrier (Figure 2.36).

Mathematically, this is described as

$$V_{out} = 0.5k \cos(2\pi(2f \pm f_m)t + \phi) \pm 0.5k \cos(2\pi f_m t - \varphi)$$

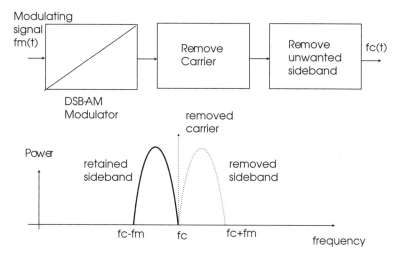

Figure 2.34   SSB modulation block diagram.

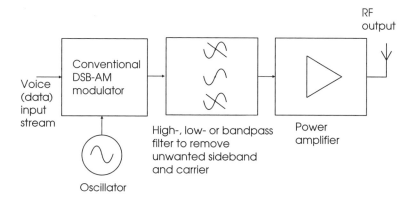

Figure 2.35   SSB indirect synthesis and filtering.

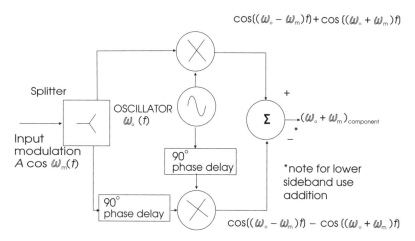

Figure 2.36   SSB direct synthesis (Hilbert modulator).

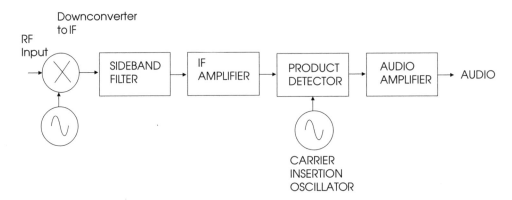

Figure 2.37   SSB demodulation using a costas loop.

The HF component is filtered out

$$V_{out} = 0.5k \cos(2\pi f_m t - \varphi)$$

$$V_{out} = 0.5k \cos(2\pi f_m)t \cos\phi \pm 0.5k \sin(2\pi f_m)t \sin\phi$$

For demodulation a costas loop is required (Figure 2.37).
  Advantages of SSB-AM:

- As already stated, all transferred power is concentrated into the 'useful' sideband carrying information. This gives transmitters efficiency theoretically upto 100 %; in reality a more practical figure is about 70 %.
- It is twice as bandwidth efficient as the DSB-AM system.

  Disadvantages of SSB-AM:

- More sophisticated demodulation is required (coherent detector with PLL, costas loop).
- It can be prone to phase distortions (due to Tx and Rx oscillators beating and Φ changing), so high oscillator specification is required at both ends. Not really heard by listeners (quality not high).

2.8.3.4   *The Aeronautical HF System and Other SSB Systems*

These use SSB modulation for reasons of spectral efficiency and a limited resource. This will be discussed further in Chapter 5. Also a number of earlier military systems are known to have used SSB as a basis.

2.8.3.5   *Suppressed Carrier Double Side Band AM*

Worth mentioning in passing, there is an intermediate version of modulation called suppressed carrier double side band AM, which lies between DSB-AM and SSB-AM. In its basic form it is the same as DSB-AM, with the carrier filtered out. It has the advantage of concentrating more power into the useful sidebands, so there is an SNR improvement; however, there is no bandwidth saving and it requires a more complex synthesis. It is rarely used as most applications will do a full conversion to the SSB, where power and spectral efficiency become important and there are no mainstream applications of this in the aeronautical bands. However, there are some specialist military systems using this for aeronautical communication.

2.8.4 Frequency Modulation

Frequency modulation is when the carrier frequency is modulated by the baseband signal. FM is used for some specialist military systems, in aeronautical telemetry (covered in Chapter 6) and finally in the adjacent band to the VHF aeronautical communications for broadcasting information. It is mainly used by broadcasting in the band between 88 and 108 MHz, just below the VHF navigation and communication bands 108.000–117.975 MHz and 117.975–137.000 MHz respectively. So a clear understanding of FM is potentially useful for adjacent channel compatibility work, and this is further discussed in Chapter 12. It is synthesized using a voltage controlled oscillator (Figure 2.38).

FM modulation is used by entertainment systems on board aircraft where channels are frequency division multiplexed (FDM) into time slots (see Section 2.10.1).

Historically, FM gives a better quality voice service. However, it is more susceptible to Doppler shifts which, even although these have been shown to be very minor in a previous section, make it impractical for mobile aeronautical communications.

2.8.4.1 Capture Effect (Hysteresis)

Although usually attributable to FM, capture effect is also relevant to AM and most other modulation types. This is a phenomenon whereby a radio receiver 'locks on' to a received signal. A higher receive power is needed to make the initial lock on than the receive power when the signal is lost. It also accounts for when two signals are present in a given volume, and the receiver tends to lock onto the one with the higher received signal power or field

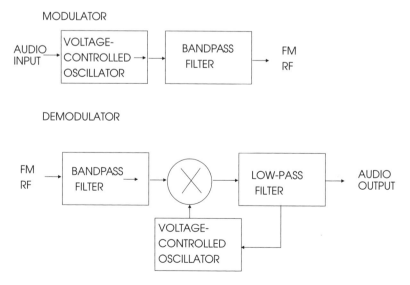

Figure 2.38 FM synthesis.

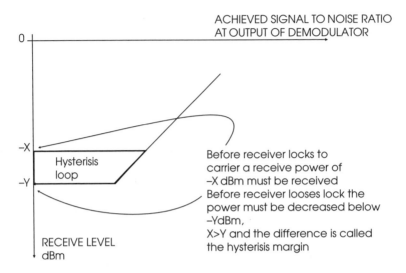

Figure 2.39　Capture effect (hysteresis).

strength; at a later time when both signal power levels are the same, the receiver will stay with the original signal. Hysteresis is important to engineering considerations (e.g. some DSB-AM receivers, extended coverage or CLIMAX as will be discussed in Chapter 3 and GPS receivers) (Figure 2.39).

2.8.5　Digital Modulation

So far the discussion has just looked at modulating analogue signals onto radio waves. However, today and increasingly in the future, there is a requirement to carry digital signals (and ever increasing data speeds) across radio. For this D/A conversion is required. Although digital modulation implies the digital domain, remember that it is actually a D/A conversion and carried as an analogue radio signal as previously discussed.

2.8.5.1　Amplitude Shift Keying (ASK)

Amplitude shift keying is when, for example, a digital 1 turns on the modulation of a fixed carrier frequency and a digital 0 signal turns it off (Figure 2.40).

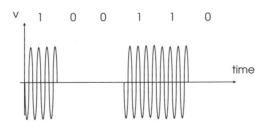

Figure 2.40　Amplitude shift keying (ASK).

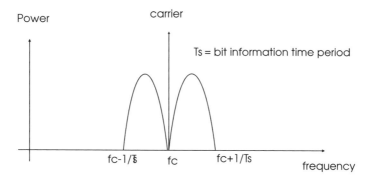

Figure 2.41 ASK spectrum components.

ASK is rarely used in aviation (except for some of the proprietary digital data systems over VHF). The main reason is to do with setting the amplitude detection threshold between what constitutes a 1 and what constitutes a 0. Consider a transmission line with some attenuation. As attenuation increases, at some point it will become impossible to differentiate between a 1 and a 0 or the thresholds will have to be set for a function of transmission line which is unnecessary when other systems can overcome this. For such a system the receiver would have to be set according to the length of transmission line. This is impractical, particularly when other modulation schemes do not have this problem.

ASK can be considered analogously with analogue AM. The spectrum of ASK (sometimes called binary ASK) is double the raw data speed in bits per second, or in this case symbols per second (baud). The RF bandwidth required is double the raw data speed in baud (i.e. 9.6 kbps will require 19.2 kHz) (Figure 2.41).

2.8.5.2 Amplitude Modulated Minimum Shift Keying (AM–MSK)

AM-MSK is a proprietary modulation scheme, using two tones f_1 and f_2; f_1 indicates a bit change from the previous bit and f_2 indicates no bit change from the previous bit. This modulation scheme is used in ACARS and VDL1 (Figures 2.42 and 2.43).

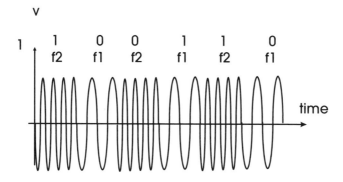

Figure 2.42 AM-MSK.

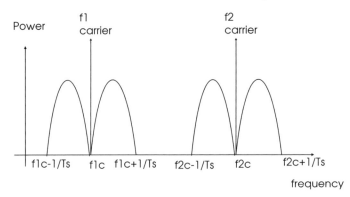

Figure 2.43   AM-MSK spectrum components.

2.8.5.3   Baud/Bit Rate and 'M-ary' ASK

It is worth making the distinction between bit rate and symbol rate. In a binary system, i.e. 2 symbol states, the bit rate is the same as the symbol rate in baud. For a 2^M system (where M is greater than 2), there are 2^M signalling states and the throughput in data is increased by a factor of 2^M for a fixed symbol speed. Table 2.5 summarizes the relationship between baud, number of signalling states and bit rate.

M-ary is another way of saying the modulation scheme is based on 2^M signalling states; i.e. for $M = 4$ ASK, there are four voltage states representing 00, 01, 10 and 11.

For a given symbol rate (i.e. number of symbols per second) the raw bit rate would be 2^M times this. However, the ability to decode the states becomes more difficult and the problem described above is exacerbated. Multisymbol ASK is not really used often in radio systems or in telecommunications line equipment.

2.8.5.4   Bipolar and Differential

The two last concepts to introduce are bipolar and differential.

Bipolar is simply a method used to say a signal state 0, for example, is represented by a voltage v and a signal state 1 is represented by $-v$. For a balanced system, it does away with a DC voltage bias that charges conductors, which can cause other problems or a slow charge up of voltage and error (Figure 2.44).

Table 2.5   Relationship between baud, bits per second and signalling states.

M	2^M	Number of discreet signalling states	Symbol rate (symbols per second or baud)	Bit rate (bps)
Binary ($M = 1$)	2	2	y baud	y
2	4	4	y baud	$2y$
3	8	8	y baud	$4y$
M	2^M	2^M	y baud	$2^M y$

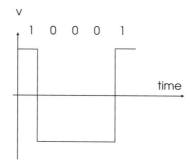

Figure 2.44 Bipolar signal.

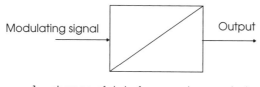

1 = change of state from previous symbol
0 = no change of state from previous signal

Figure 2.45 Differential signalling.

Differential is when one amplitude represents a change of state between a digital value and the preceding value, and another state represents no change in digital value. Differential encoding has the advantage that clock synchronization is integrally embedded in the signal in which a received signal can recover the clock from the received data stream. It can also do away with timing errors from non-synchronized transmitter and receiver clocks (Figure 2.45).

2.8.5.5 Frequency Shift Keying

In the binary FSK modulator, a binary 1 input will modulate a frequency f_1 and a binary 0 state will modulate a frequency f_2 (Figure 2.46). There is an analogy here with FM. Bipolar Frequency Shift Keying is when a zero is represented by one frequency and a 1 is represented by another frequency. Bipolar frequency shift keying (BFSK) is the bipolar version of this (Figures 2.47–2.49).

The spectrum bandwidth occupied by BFSK $= 2B(1 + \beta)$, where B is $1/t_1$, the symbol rate of the data signal and β is $\Delta f/B$. (There is no point in increasing the modulation index beyond 1; $\Delta f = B$.) At this point, BFSK is four times the data rate. As can be seen, FSK has a wide RF spectrum.

2.8.5.6 Phase Shift Keying

PSK is when the carrier phase is changed to represent the symbol wanted. The scheme for 4PSK is shown in Figure 2.50, together with what is called a constellation or phase diagram. PSK can also be differential (denoted by DPSK); it can also be M-ary, i.e. 4PSK (sometimes called quadrature PSK) or 8DPSK; and the nomenclature builds up (Figures 2.51 and 2.52).

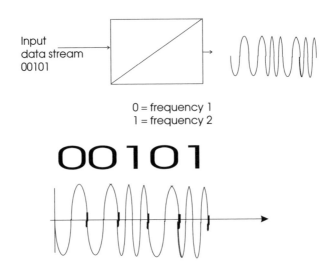

Figure 2.46 Frequency shift keying.

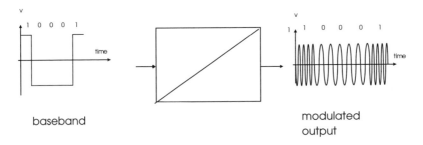

Figure 2.47 Bipolar FSK.

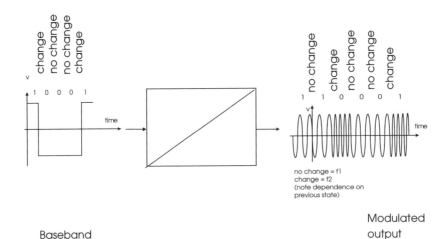

Figure 2.48 Bipolar differential FSK.

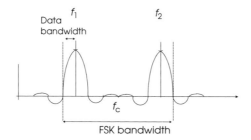

FSK is sum of spectra of two ASK waves
FSK bandwidth = $(f_1 - f_2)$ + 2BW (data baseband)

Figure 2.49 FSK spectrum components.

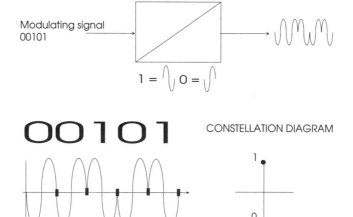

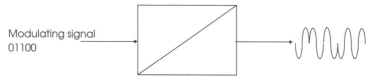

Figure 2.50 Phase shift keying.

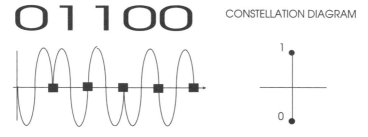

Figure 2.51 DPSK.

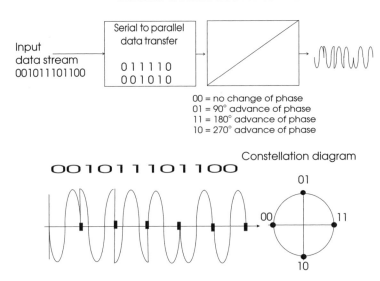

Input data stream 001011101100

Serial to parallel data transfer

011110
001010

00 = no change of phase
01 = 90° advance of phase
11 = 180° advance of phase
10 = 270° advance of phase

Constellation diagram

001011101100

01

00   11

10

Figure 2.52 4DPSK.

2.8.5.6.1 Quadrature Phase Shift Keying (QPSK)

It offers the following features:

- baud rate stays the same;
- signalling states double;
- bit rate doubles;
- RF bandwidth remains the same;
- required SNR goes up for detection.

2.8.5.6.2 Eight-Phase Differential Shift Keying (8DPSK)
(Figures 2.53 and 2.54)

It offers the following features:

- baud rate stays the same;
- signalling state quadruples;
- bit rate quadruples;
- RF bandwidth remains the same;
- required SNR goes up for detection.

Obviously as the modulation index (M) goes up, the difference in voltage and phase between the different signalling states is decreased; hence from a practical viewpoint, it is harder to distinguish between the states, and as such as the M-ary level is raised, the required SNR in the analogue environment necessary for successful demodulation goes up. This is best illustrated by Figure 2.55.

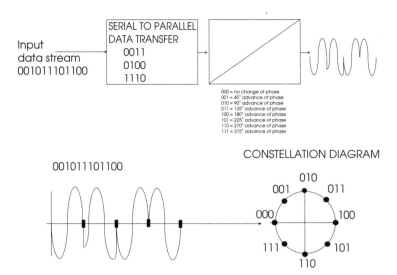

Figure 2.53   8DPSK.

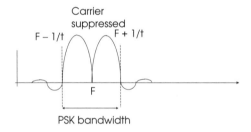

Figure 2.54   Spectral components of the various PSK family.

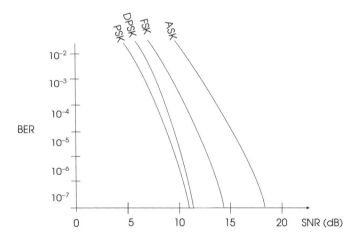

Figure 2.55   Signal to noise ratio (SNR) versus bit error ratio (BER) for the PSK family.

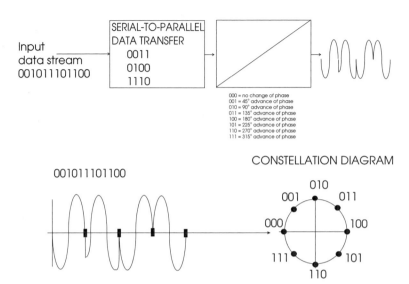

Figure 2.56   8DPSK synthesis.

8DPSK modulation is used for a number of VHF datalink systems (notably VDL2 and VDL3) described in Chapter 3. Thus let us take a closer look at this modulation specifically (see Figure 2.56).

The input serial bitstream is scrambled. This is to give better spectrum shaping independent of $1 + 0$ content. It is then serial to parallel converted into octet sets. (Each octet is represented in modulation as one of eight possible phase changes selected for transmission.) The final signal is shaped using a raised cosine filter (to reduce spectral side lobes and energy that spills into adjacent channel). The advantage of going to multilevel PSK is that the data throughput goes up, but the used RF bandwidth remains steady. Obviously the higher the M-ary modulation, the more sensitive a system will be to phase distortions, and for a given symbol rate, the required SNR will go up to maintain a constant BER (bit error rate).

PSK gives a tighter spectral mask than FSK and ASK. This is usually the reason it why is more common to see PSK-based modulation schemes. They are also not sensitive to amplitude distortions, but only to phase distortions.

2.8.5.7   Quadrature Amplitude Modulation (QAM) and Trellis Code Modulation (TCM)

QAM is a combination of the phase domain. The number of possible states = 360/number of phase changes used. The modulator block diagram is shown in Figure 2.57. The demodulation process for this becomes quite complex, but is not unachievable with modern technology. It is used frequently in high-capacity data point-to-point radio links. 64QAM, 128QAM, 256QAM and even 518QAM are not uncommon. Obviously, these have high spectral efficiency in the amount of data per unit RF bandwidth they are able to pass. But consequently a very good SNR is needed to retain a high BER.

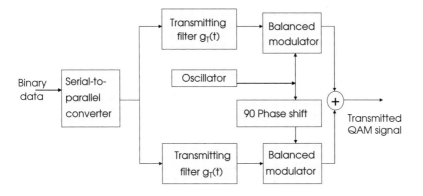

Functional block diagram of modulator for QAM

Quadrature Amplitude demodulation

Demodulation and detection of QAM signals

Figure 2.57   Quadrature amplitude modulation/demodulation.

2.8.5.8   *Trellis Code Modulation* (Figure 2.58)

Initially the modulator states are based on a trellis shown in Figure 2.58. This scheme is intelligent and involves more signal processing theory than the simpler prior modulation schemes. It works on the principle of putting states that are likely to occur simultaneously as far apart on the constellation diagram with maximum distance of separation (sometimes called eclusian distance). Similarly a binary sequence that is unlikely to occur can be put together on the constellation diagram with minimum eclusian distance. This offers an effective coding gain over the simpler QAM and for a given BER and data throughput would consequently require a lower SNR than QAM (Figure 2.59).

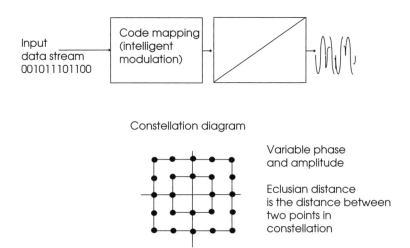

Figure 2.58   Trellis code modulation (TCM).

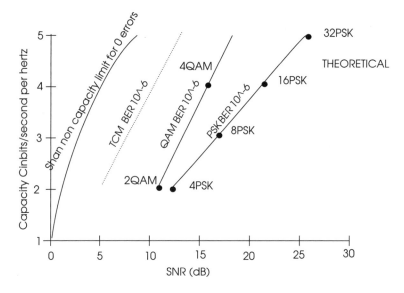

Figure 2.59   Comparison between QAM and TCM SNR versus BER.

2.8.5.9 Gaussian Frequency Shift Keying (GFSK)

In GFSK, a bipolar or non-return to zero signal is applied to FSK modulator and the side lobe levels of the spectrum are further reduced by passing the modulated waveform through a pre-modulation Gaussian pulse shaping filter (Figure 2.60). It offers the following features:

- attractive for its excellent power efficiency (due to the constant envelope); where power is more evenly spread around the band.
- baud rate stays the same;

GAUSSIAN PHASE SHIFT KEYING

NRZ data → Gaussian Low-pass filter* → FM transmitter → GFSK

*This filters out sharp transitions and hence keeps spectral components in the band considered and spreads the spectral power density more evenly. This increases spectral efficiency.

Figure 2.60   Gaussian frequency shift keying (GFSK).

- signalling states go up;
- bit rate goes up;
- RF bandwidth remains the same;
- required SNR goes down for detection.

GFSK has the advantage that its spectral shape is more uniform than its PSK or FSK equivalents. This gives it a further BER versus SNR advantage (Figures 2.61 and 2.62). It is the modulation chosen for the latest VDL mode radio, VDL Mode 4.

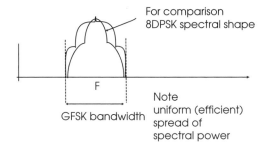

For comparison
8DPSK spectral shape

F

GFSK bandwidth

Note
uniform (efficient)
spread of
spectral power

Figure 2.61   GFSK spectrum components.

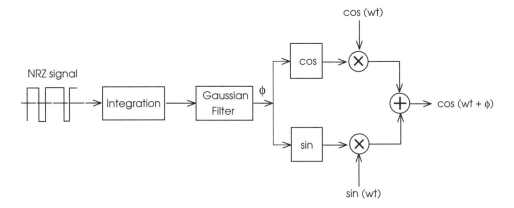

Figure 2.62   GFSK synthesis.

Table 2.6　Shannon's law and relationship between M, SNR and spectral efficiency for errorless transmission.

M	2^M	Minimum S/N required	Maximum spectrum efficiency (maximum capacity)
1	2	3	1 bits/Hz
2	4	15	2 bits/Hz
3	8	63	4 bits/Hz

2.9 Shannon's Theory

Shannon's theory provides a theoretical relationship between bits per second, baud and SNR for error-free transmission (Table 2.6).

$$R \text{ (bps)} = M \text{ (number of signalling states)} \times r \text{ (baud or signalling rate)} \quad (2.31)$$

$$C = B \, \log_2 \left(1 + \frac{S}{N} \right) \quad (2.32)$$

where C is channel capacity in bps, B is bandwidth in Hz, S/N is SNR, R is the data rate and is always $< C$ for errorless transmission. If $R > C$, errors will occur. Theoretical relationship between M and SNR:

$$M = \left(1 + \frac{S}{N} \right)^{0.5} \quad (2.33)$$

Or

$$\frac{S}{N} = M^2 - 1 \quad (2.34)$$

2.9.1 Non-Errorless Transmission

Of course Shannon's law does not apply if it is decided to push past this theoretical limit of data throughput if errors can be tolerated.

Actually it is sometimes a prudent approach to tolerate the errored environment and to make facilities for correcting them (by forward error correction (FEC) and cyclic redundancy coding (CRC)); of course FEC and CRC too are overhead and not useful data payload. However, this can be a more efficient way to operate and maximize data throughput. Alternatively the errors can just be tolerated and the application layer of the data system can be designed to tolerate this or to correct it.

Also in a number of digital systems, errors will be present from the propagation problems already present and these have to be tolerated. These too can be corrected by FEC or CRC checks, or other mechanisms discussed later in Section 2.12.

2.10 Multiplexing and Trunking

Multiplexing is the concept of putting multiple channels into one radio channel resource or down one cable or one fibre.

2.10.1 Frequency Division Multiplexing (FDM)

FDM is when a number of different channels are available (this can be a radio resource or a cable resource) (Figure 2.63).

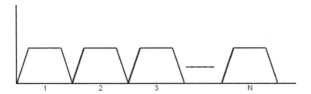

Figure 2.63 Frequency division multiplexing (FDM).

2.10.2 Trunking

Trunking is when one channel resource is shared between many users.

2.10.2.1 Example

From Figure 2.64.

- Consider a bank with four tellers in operation, each with individual queues.
- Now consider a bank with four tellers in operation, but one queue feeding to each of the four tellers as they become available.

This later version is trunking and shows decreased wait times for the average queue and efficiency in how each teller (or channel in RF terms) is used. It also shows how a relatively scarce resource can be optimally shared.

Such topologies are used frequently in radio systems too, more specifically RF radio channels.

The advantage of non-trunked working can be that all the users are aware of the activity of that user group. It works in a 'broadcast mode', sometimes called 'talk around' which is quite often the case for systems deployed in operational environments such as the ATC landing radio system and oil industry mobile communications. The disadvantage of such a topology is, it is potentially an inefficient use of resources, which can cause congestion. Trunking can enable the RF spectrum requirement to be significantly compressed.

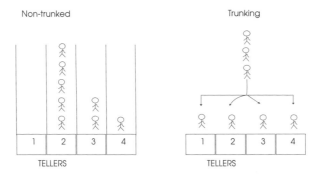

Figure 2.64 Trunking and non-trunked.

To put this in perspective, consider the use of a number of radio channels in both modes and consider how many radios they can support. This is best described by Erlang B Traffic Table.

$$A = yS \qquad (2.35)$$

where A is traffic in erlangs, y is mean call arrival rate (calls per unit time) and S is mean call holding time (same units as y).

Erlang B Traffic Table

Erlang B is a model in which all blocked calls are lost. This is the scenario for mobile communication systems and telephony. When the caller receives a busy signal, the call is lost and the caller must redial repeatedly until resource becomes available.

Erlang B calculates the blocked call probability for a given traffic load and a number of resources. $P_B(N,A)$ is the probability that a caller will receive a busy signal with a traffic load of A erlangs and N available lines.

$$P_E(N,A) = \frac{A^N/N!}{\sum_{i=0}^{N} A^i/i!}$$

Tables shows A, offered load in erlangs for a given number of lines (N) and a given blocking probability (P_B) in percentage.

Offered Erlangs

Table 2.7   Erlang-B blocking probabilities.

Number of lines	Blocking probabilities (%)							
	0.01	0.05	0.1	0.5	1	2	5	10
1	0.0001	0.005	0.001	0.005	0.0101	0.0204	0.0526	0.1111
2	0.0142	0.0321	0.458	0.1054	0.1526	0.2235	0.3813	0.5954
3	0.0868	0.1517	0.1938	0.3490	0.4555	0.6022	0.8994	1.271
4	0.2347	0.3624	0.4393	0.7012	0.8694	1.092	1.525	2.045
5	4520	0.6486	0.7621	1.132	1.361	1.657	2.219	2.881
6	0.7282	0.9957	1.146	1.622	1.909	2.276	2.96	3.758
7	1.054	1.392	1.579	2.158	2.501	2.935	3.738	4.666
8	1.422	1.83	2.051	2.73	3.128	3.627	4.543	5.597
9	1.826	2.302	2.558	3.333	3.783	4.345	5.37	6.546
10	2.26	2.803	3.092	3.961	4.461	5.084	6.216	7.511
11	2.722	3.329	3.651	4.610	5.16	5.842	7.076	8.487
12	3.207	3.878	4.231	5.279	5.876	6.615	7.95	9.474
13	3.713	4.447	4.831	5.964	6.607	7.402	8.835	10.47
14	4.239	5.032	5.446	6.663	7.352	8.2	9.73	11.47
15	4.781	5.634	6.077	7.376	8.108	9.01	10.63	12.48
16	5.339	6.25	6.772	8.1	8.875	9.828	11.54	13.5
17	5.911	6.878	7.378	8.834	9.652	10.66	12.46	14.52
18	6.496	7.519	8.046	9.578	10.44	11.49	13.39	15.55
19	7.093	8.17	8.724	10.33	11.23	12.33	14.32	16.58
20	7.701	8.831	9.412	11.09	12.03	13.18	15.25	17.61

Usage of a radio channel is dependent on

- number of mobiles using it;
- typical erlang demand per mobile;
- probability of blocking required.

It is assumed that number of sources $\gg$ number of channels (N) and rate of call arrival is constant.

Table 2.7 shows that the probability of call being blocked is a function of

- average erlang traffic per user;
- number of users;
- number of channels available.

This demonstrates the probability of a channel being busy (unobtainable) when used by n users, each with an average requirement of E (in erlangs) loading.

Ultimately, the decision whether to use trunked or non-trunked topologies depends on many things, including safety and the minimum delay acceptable under the quality parameter of the three C's (described later in Chapter 3).

2.10.3   Time Division Multiplexing (TDM)

This is when there is one RF channel shared in time between the different users. If a suitable TDM repetition frequency is chosen, the user does not need to be aware of the discontinuity (Figure 2.65). This is the principle deployed in modern ground cable and fibre networks to which the radio systems described in this book need to interface.

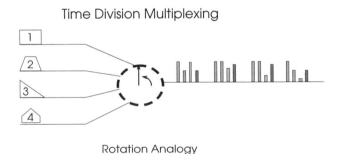

Figure 2.65   Time division multiplexing (TDM).

2.10.4   Orthogonal Frequency Division Multiplexing (OFDM) and Coded
   ### OFDM (Figure 2.66)

OFDM distributes the data over a large number of carriers that are spaced apart at integer frequencies. This spacing provides the 'orthogonality' in this technique, which prevents the demodulators from seeing frequencies other than their own. OFDM offers high spectral efficiency, resilience to RF interference and high immunity to multipath distortion effects.

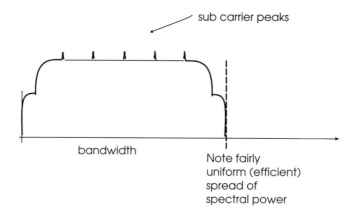

sub carrier peaks

bandwidth

Note fairly
uniform (efficient)
spread of
spectral power

Figure 2.66   Orthogonal frequency division multiplexing (OFDM) and coded OFDM.

OFDM is sometimes called multicarrier or discrete multitone modulation. It is the modulation technique used for digital TV in Europe and also for the IEEE 802.xx standards. Coded OFDM is a variant of this that effectively spreads the spectrum.

2.11   Access Schemes

Extending the FDM and TDM concepts to the domain of digital radio introduces a subtlety called access, the concept of a non-continuous use of a time slot of frequency channel.

2.11.1   Frequency Division Multiple Access (FDMA)

This is when a radio resource is shared in an FDM manner. Any user may use one of the frequencies for a given amount of time. A call may be initiated and continued for a variable length of time. This is a form of trunking (Figure 2.67).

FDMA is a method used for many mobile radio applications. The first generation of mobile phones (AMPs and TAC) used this method as well as some of the first and second generation

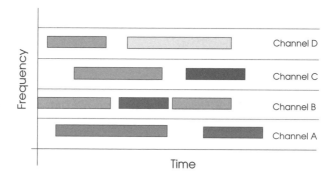

Channel D

Channel C

Channel B

Channel A

Frequency

Time

Figure 2.67   Frequency division multiple access (FDMA).

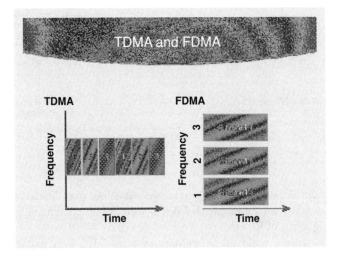

Figure 2.68   Time division multiple access (TDMA) and frequency division multiple access (FDMA).

of mobile satellite demand assigned multiple access or DAMA systems. The channels are generally analogue channels. Arguably the VHF AM(R)S communication system is a form of FDMA – in fact it is actually a simplex/broadcasting mode version of this.

2.11.2   Time Division Multiple Access (TDMA)

This is an extension of the TDM principle already discussed in the radio environment (see Section 2.10.3; Figure 2.68). It exploits the use of dividing the time domain between the users. This technology is nearly always digital and is sometimes referred to as second-generation technology or second-generation digital mobile radio technology.

TDMA is used by technologies such as GSM, PCS 1800, DECT, TETRA, digital satellite mobile and also digital DAMA. It gives an improved spectral efficiency over the previous FDMA analogue technology. Other advantages include security encryption functions and a greater agility for services.

2.11.3   Code Division Multiple Access (CDMA)

Having considered cutting up the frequency domain (FDMA) and cutting up the time domain (TDMA), now consider using these two domains continuously (i.e. transmitting across all the available bandwidth and continuously intime) and at the same time distinguishing between each of the different mobiles using the spectrum via the unique codes they are individually assigned. This, in simplicity, is CDMA (Figure 2.69).

Consider a user data stream that is multiplied by a unique pseudo-random bitstream to produce pseudo-noise-like modulation. When this same noise-like data stream is multiplied by the same pseudo-random bitstream, the original data stream is recovered. This process is sometimes called spread spectrum and is a process whereby the data is spread right across the available RF spectrum channel available. In addition, it is not so discernable from the noise in

Figure 2.69   Code division multiple access (CDMA).

which it sits. In fact it is actually possible to lift the signal out from 'under' the noise. That is the combined transmission by multiple mobiles is seen as noise and an individual channel is underneath this (in terms of RF power).

This method is widely deployed by military systems (also with frequency hopping included) as well as by the next generation of digital mobile telephones, sometimes called third-generation or 3G. It is also deployed by navigation satellites such as GPS and Galileo (Figures 2.70–2.72).

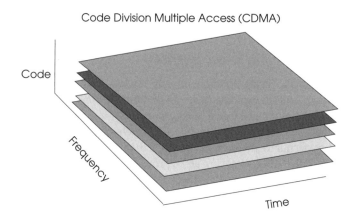

Figure 2.70   The CDMA domain, time versus frequency versus code.

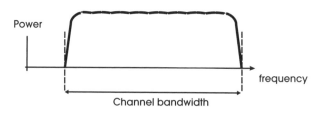

Figure 2.71   The CDMA spectrum, frequency domain.

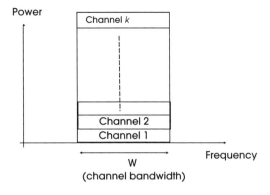

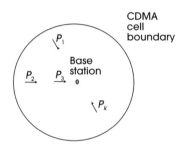

Figure 2.72 The CDMA cell.

2.11.3.1 CDMA Principles

Consider a CDMA system as described in Figures 2.72 and 2.73. Assume all users transmit with the same frequency in the same RF bandwidth, and codes are used to discriminate between wanted and unwanted signal. Now consider a situation where there are k mobile stations in a given cell as shown.

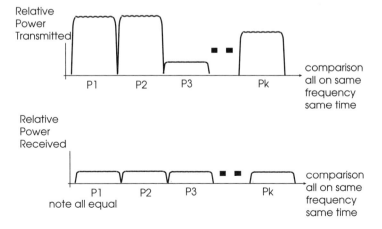

Figure 2.73 CDMA balanced network.

In order to optimize the system, the receive power at the base station is controlled (by the base station control channel), so that the power levels of each of the incoming mobiles are equal. This is a very important requirement for CDMA working (i.e. the system is balanced).

The same principle is also deployed in the reverse uplink (base station to mobile) direction, but for now consider only the mobile to base station direction. Consider one of the receive paths. If C is equal to wanted receive signal power from that mobile at the base station (in dBW), then k is the total number of mobiles in the cell and I is equal to total interference power coming into the local base station. For this balanced situation to occur,

$$I = (k - 1)C$$
$$\frac{C}{I} = 1/(k - 1)(\text{dB}) \tag{2.36}$$

Now if I_0 is the interference power density in dBW/Hz and W is the signal bandwidth in Hz, then

$$I_0 = I/W(\text{dBW/Hz}) \tag{2.37}$$

Similarly, if R is the data rate of the signal in bits per second.

the energy per bit $E_b = C/R$ joules $\tag{2.38}$

Linking all three of these equations together,

$$k - 1 = I/C = I_0W/C = (W/R)/((C/R)/I_0)) = (W/R)/(E_b/I_0)$$

This gives us the basic set of CDMA equations governing its operation. W/R is sometimes called the processing gain or spreading factor.

For large values of k, $k \approx (W/R)/(E_b/I_0)$ $\tag{2.39}$

(W/R typically ranges from 4 to 256).

Coding gain $= 10 \log$ spreading factor $\tag{2.40}$

2.11.3.2 Frequency Domain Duplex (FDD) and Time Domain Duplex (TDD)

For a duplex system, there are two methods by which the go-and-return path can be structured. The first is in the frequency domain, where a channel each is dedicated to up- and downlinks. This is called frequency division duplexing (Figure 2.74). The second is in the time domain, where each transmit and receive path takes it in turn to use the resource in a time domain duplex (TDD) mode (Figure 2.75).

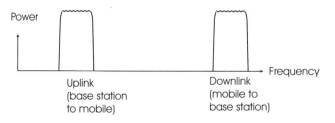

Figure 2.74 Frequency domain duplex (FDD) operation.

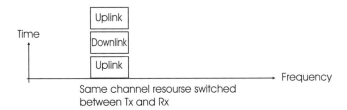

Figure 2.75 Time domain duplex (TDD) operation.

2.11.3.3 CDMA Applications

As discussed earlier, CDMA is sometimes called the third generation of digital mobile radio. It has been developed primarily for military systems and also for public mobile networks. In this second area of development, a number of *de facto* standards have emerged. The two principal variants are CDMA 2000 and IS 95.

CDMA 2000 is sometimes called *IMT 2000*. It uses a channelization of 5 MHz (WCDMA) or $N \times 1.25$ MHz and is deployed generally as TDD, but can also be FDD. *IS 95* was the first type of 3G CDMA system proposed (sometimes called CDMA one) and uses a channelization bandwidth of 1.25 MHz in FDD mode. Other variants have emerged, including some proprietary Chinese standards, but are not discussed further. These standards become relevant because they are a potential option for adapting to make a new aeronautical mobile standard.

The advantage of using FDD is that there are no cell range limitations as in the case for TDD or TDMA structures, while the advantage of using a TDD topology is that only one RF spectrum channel allocation would be required to satisfy both the up and down link parts.

2.12 Mitigation Techniques for Fading and Multipath

In this section, techniques are identified that are commonly used in radio receivers to improve availability and reliability from anomalies encountered with propagation. These techniques include the following.

2.12.1 Equalization

For any radio link, some baseband frequency components experience more attenuation than others, which is extenuated when fading occurs, particularly multipath fading. Here it is found that some frequency components of the baseband experience rapid fluctuations in attenuation, sometimes called fast fading or selective fading (Figure 2.76). Thus at the receiver some notches are present in the received spectrum and correction is required to compensate for this factor so that the data stream can be successfully reconstructed.

There are numerous equalization schemes used for compensating the variation in attenuation with frequency response. In analogue radio systems the IF portion of the radio can be set up to compensate for this. (This is the long-term correction.) For digital radio it is found that the fluctuations are dynamic in nature. Consequently, often a known word is sent in the data stream. This known word will arrive with errors and when demodulated, the signature of the fading environment can be calculated, and accordingly the equalization required to correct

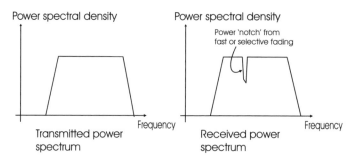

In the time domain, there will be bit errors for a small duration correlating to when fade starts and finishes.

Figure 2.76   Selective fading.

this errored word back to its former unerrored state. This differential or feedback equalization can be applied to the rest of the raw data before the FEC and block coding algorithms. It is sometimes called transversal equalization and is common in high-capacity microwave radio systems and mobile systems (Figure 2.77).

2.12.2   Forward Error Correction and Cyclic Redundancy Checking

This is where a known amount of bits is set aside for error detecting and correcting of a data stream. With FEC, there are two stages to the process: The code enables a certain amount of errors to be detected and usually to a lesser degree a number of errors that can be corrected for any given block size.

Convolutional coding is usually an extended version of FEC, involving more overhead but an even better ability to detect and correct errors. It is an extremely powerful tool. This can be put as a frame buffer or in some of the redundant end bits of the data frame.

This uses polynomial algorithms for optimizing the coding correction ability on a block of data. It should be noted that with FEC and block coding, potentially the data throughput of a system is reduced; however, the BER is usually improved, and the trade-off is an engineering optimization (Figure 2.78). This gives a very basic introduction to signal processing and coding gain advantages. This discipline and practice, which is highly mathematical and probabilistic is becoming increasingly popular and useful with the availability and economy of increased computer processor power; however, the detailed application is outside the scope of this book.

2.12.3   Interleaving

This method primarily protects against multipath and fast fading environments. Interleaving is a process where a real-time frame is broken down and block sequences are interleaved in time (and phase on quadrature modulators). In this way in a fast fading environment, not all of a section of data will be lost but rather parts from each block of digital data. Thus, when it comes to reconstruct the information in the receiver, a part of this can be done, and with the help of the FEC, CRC coding and equalization and all is not lost. A degradation in BER quality is better than total loss (Figure 2.79).

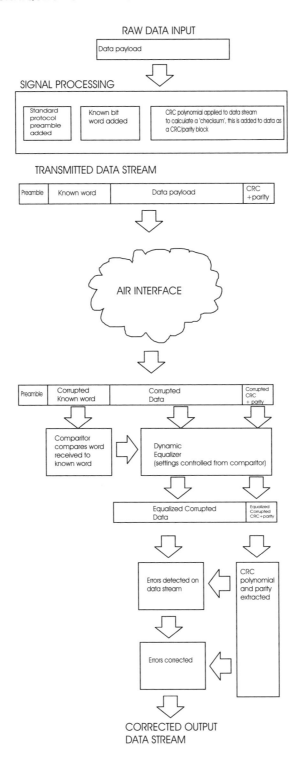

Figure 2.77   Equalization process and block diagram.

FEC

Preamble	Knownword (usuallyvery small)	Datapayload	FEC bits
a bits	b bits	c bits	d bits

note error detecting and error correcting ability is a function of the relationship between size a, b, c anddand ultimately a mathematical polynomial usually use donasynchronous data streams

CRC

start/ control frame	Datapayload	CR C
	Datapayload	CR C
	Datapayload	CR C
	Datapayload	CR C

note CRC has a similar function and is also described by apolynomial mathematically, in practice it is usually deployed on larger data blocks (synchronously)

Figure 2.78   FEC and CRC.

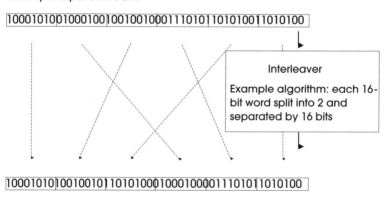

Example Input Bitstream

`1000101001000100100100100011101011010100110101001`

Interleaver

Example algorithm: each 16-bit word split into 2 and separated by 16 bits

`10001010100100101110101000100010001110101110101100`

Example Output Bitstream

Figure 2.79   The process of interleaving.

2.12.4   Space Diversity

This is when two or more receive antennas are used, separated in space, using the unlikelihood that fading or multipath will occur to two geometric receive points simultaneously. In fact, there is a statistical relationship between the spatial separation of two receiving antennas and the likelihood of both receiving faded signals simultaneously.

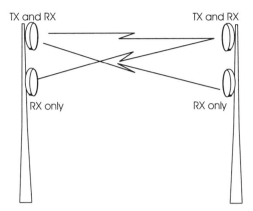

Figure 2.80 Space diversity topology.

There is a formula called the space diversity improvement factor, denoted I, which is a function of the wavelengths between the two receiving antennas, path length, frequency and the type of combining employed, (be it baseband or IF combining.) This is not discussed further here but is the subject of much study within ITU-R – particularly the study group defining the propagation studies (Figure 2.80). (See ITU-R. P recommendations).

Space diversity combining can be done in the IF stages of the amplifier in the analogue domain (called IF combining) or alternatively, a quality detection circuit can switch the better signal path in or out (called baseband switching) (Figure 2.81). It can also be considered to a degree as a form of equipment protection because it gives an increased reliability of the system (as at least the receivers must be duplicated).

This technique is used frequently for microwave point-to-point links. It is also used by passive receivers and soft handovers in the CDMA environment.

2.12.5 Frequency Diversity

Frequency diversity means sending the same information simultaneously on two or more frequency channels (say $N + 1$ systems) (Figure 2.82). This method provides protection against multipath anomalies (in that it is unlikely to occur on two separate frequencies simultaneously) and it also provides active equipment protection, although usually antennas used are the same for both systems and hence the new weak point. This technique is used frequently by terrestrial LOS systems and also satellite links, (particularly the hub station in a network) (Figure 2.83).

This principle can be extended to $N + 1$ working of a digital bearer, where N unduplicated channels basically share the last channel as a common protection channel in case one of the N channels fails (Figure 2.84).

However additional RF spectrum is the penalty required to achieve frequency diversity.

2.12.6 Passive Receiver Diversity

Another technique to mitigate against multipath problems, poor coverage or equipment reliability is the concept of placing a receiving antenna at a separate geographical location. This could be hundreds of metres away, could be kilometres away or in some cases, even over the

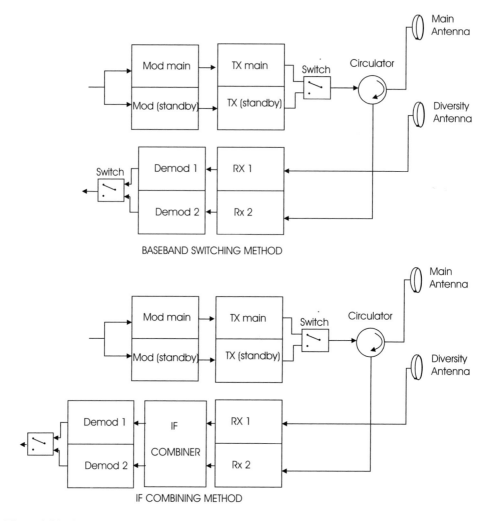

Figure 2.81 Space diversity combining.

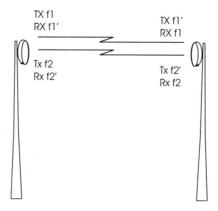

Figure 2.82 Frequency diversity.

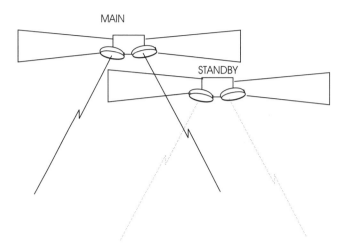

Figure 2.83 Satellite diversity.

horizon. The ability to pick up reception at multiple places, compare and correlate can provide an overall gain in availability and reliability (Figure 2.85).

This technique is exploited by the military systems, which have the added advantage of being able to receive only and hence their location is not given away to the enemy. These systems usually require phenomenal processing power to correlate receive data streams and options. It is also exploited for the legacy VHF communications system in a concept called extended coverage or climax operation. This will be discussed further in Chapter 3 of this book.

2.13 Bandwidth Normalization

For any given radio system, the transmitter is known to transmit its signal within a given RF bandwidth. The actual profile of the transmitted spectrum can be described by the spectral density of the emission masked by frequency (Figure 2.86).

As shown previously, the relative shape of the emitted spectrum is a function of modulation type but also the RF amplifier transfer function and the spectrum shaping carried out by the RF filtering just prior to the antenna.

On the receiving side, there is a similar RF receive filter characteristic. When defining a receiver bandwidth, usually the frequency range is described between the 3-dB points. (The 3-dB points are the points where the cut-off receiver filter is attenuating the incoming frequency components by a relative 3 dB when compared to receive centre frequency.) (Figure 2.87).

This 3-dB bandwidth is a useful piece of information; however, it does not give the designer or system planner all the information required. The actual receiver characteristic would be needed for this and it would show how fast the signals are attenuated when going out of band. (Out of band usually refers to when the signals are in frequency beyond the 3-dB points.)

While studying a receiver characteristic and trying to understand the effects another transmitting source (other than the one wanted) may have on the receiver under study, it may be found that the transmitter and receiver are operating in different 3-dB bandwidths. Thus in order to understand what is happening to a 'victim' receiver, the signal from the unwanted transmitter

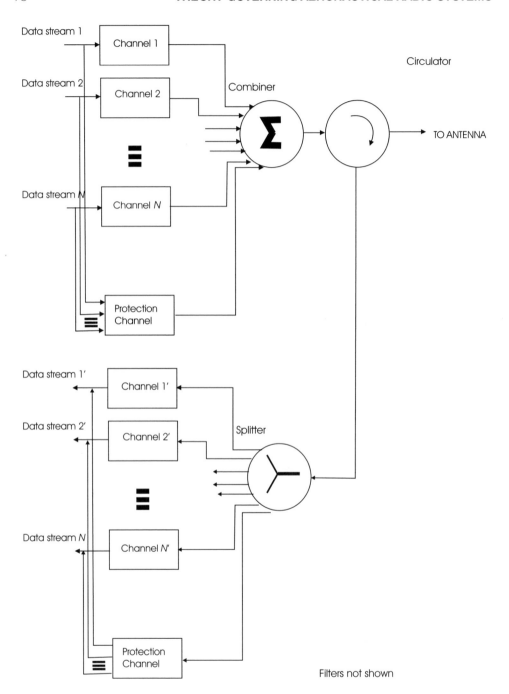

Figure 2.84   Extended principle to $N + 1$ working.

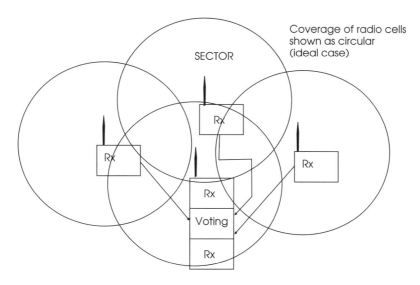

Figure 2.85 Passive receiver diversity.

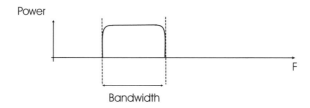

Figure 2.86 Emission spectrum density.

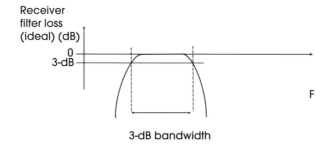

Figure 2.87 Receiver filter characteristic and 3-dB points.

needs to be redefined in the receiver bandwidth. This process is called normalization. This technique is particularly necessary for carrying out compatibility studies, interference studies, etc., when unlike signals need to be compared on a common basis (Figure 2.88).

Normalization is also a useful technique to benchmark various RF systems to a frequency. Quite often, 1 MHz is chosen as a benchmark bandwidth, also 3 kHz or 25 kHz are other commonly occurring reference bandwidths.

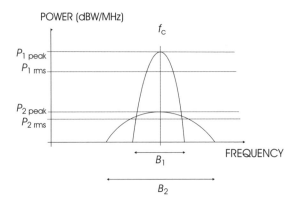

Figure 2.88   Bandwidth normalization.

The average power of the signal across the bandwidth under consideration

$$P_{1\text{rms}}(\text{dBW}/B_1 \text{ MHz}) = \frac{1}{B_1} \int_{f_{\text{low1}}}^{f_{\text{high1}}} \text{Power } \mathrm{d}f$$

The signal to be compared with it similarly has a power

$$P_{2\text{rms}}(\text{dBW}/B_2 \text{ MHz}) = \frac{1}{B_2} \int_{f_{\text{low2}}}^{f_{\text{high2}}} \text{Power } \mathrm{d}f$$

The areas under each of the signals must be the same.
    Therefore, to convert powers

$$P_2(\text{dBW}/B_2\text{MHz}) = P_1(\text{dBW}/B_1 \text{ MHz}) + 10 \log B_2 - 10 \log B_1 \qquad (2.41)$$

Note for a fuller discussion on this and a convolutional approach see ITU-R recommendations.

2.14   Antenna Gain

2.14.1   Ideal Isotropic Antenna

To recap from Section 2.4, remember the concept of the isotropic antenna as being one with the following properties:

- uniformly radiating in all directions;
- pinpoint source of no volume;
- a reference antenna;
- ideal (cannot exist);
- 0-dBi gain by reference or definition.

2.14.2 Practical Realizations

In practice, all antennas have a radiation pattern envelope (RPE), which is defined as the gain geometry of an antenna relative to the isotropic ideal; i.e. an antenna that has more gain in some directions than others. The RPE can be presented as a polar pattern of gain versus angle for the azimuth direction (horizontal scanning through the 360°) and the elevation aspect (scanning 360° vertically about the antenna centre point) (Figure 2.89).

Sometimes an all-round (as close to isotropic as possible) or omnidirectional RPE is preferred to give good overall radio coverage (this is a typical mobile application). Or sometimes it suits the design engineer to have a focused antenna: focusing the RF energy in a convenient direction or equally importantly, rejecting RF energy coming form an unwanted angle to the antenna (see Figures 2.95 and 2.97). This is often the case for point-to-point communications or communications where the RF spectrum is highly utilized or shared and the RF resource can be considered 'at a premium'.

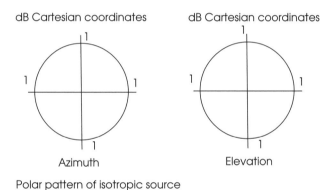

Figure 2.89 Isotropic azimuth and elevation RPEs.

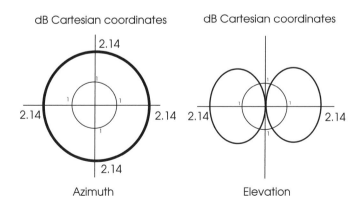

Figure 2.90 Typical omnidirectional antenna pattern, the closest practical realization of the ideal isotropic antenna.

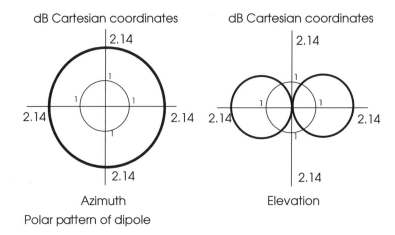

dB Cartesian coordinates   dB Cartesian coordinates

Azimuth   Elevation

Polar pattern of dipole

Figure 2.91   The dipole and its RPE.

2.14.3   *Some Common Antennas Used for Aeronautical Communications*

2.14.3.1   *The Dipole*

Probably the most basic type of antennas, the dipole has two terminals or 'poles' (Figure 2.91).

For resonance the conductor is an odd number of $\lambda/2$ in length (1, 3, 5, etc.), but usually just one $\lambda/2$. It is centre-fed, at the centre of the standing wave voltage.

Centre current is maximum where the voltage is at a minimum. It has a low impedance feed point 73.13 Ω, which is easy to match to 75-Ω feeder coaxial cable.

The feed impedance can be changed by introducing a number of proximate metal elements; this changes matching and VSWR.

It is slightly shorter than $\lambda/2$, with antenna gain of 2.14 dBi ($= 0$ dBd).

Its polarity is dependent on the direction of orientation. It is usually used in vertical mode.

2.14.3.2   *The Folded Dipole*

This has its conductors folded back 'like a coat hanger'.

Its electrical properties give it a fourfold increase in feed impedance to around 300 Ω. Thus it is less prone to impedance variations than the dipole.

It is $\lambda/2$ in length, fed at the midpoint and has a wider RF bandwidth of use.

It has the same radiation pattern as a dipole antenna.

Again the polarity is a function of the oriented direction. It is usually used in vertical mode (Figure 2.92).

2.14.3.3   *Quarter-Wave Vertical Antenna*

This antenna is used regularly from MF to VHF.

It is a quarter wave or quarter wave above a ground plane antenna.

It is omnidirectional in the horizontal plane. It consists of a single end-fed element. In elevation, with most energy concentrated on horizontal 'lobe', making it very suitable for mobile type applications. It has a low impedance around 20 Ω.

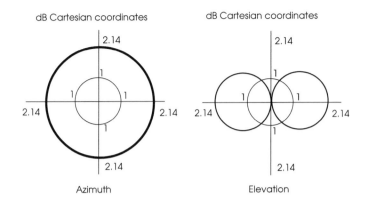

Figure 2.92 The folded dipole and its associated RPE.

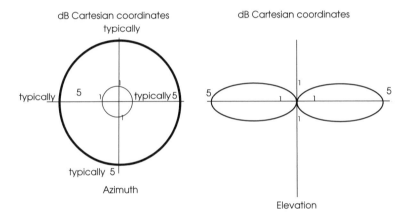

Figure 2.93 Quarter-wave vertical antenna and associated RPE pattern.

The grounding is important (RF mat at tower or large metal body, car or aircraft).

Alternatively quarter-wave horizontal radials can be used to simulate the ground plane.

If radials are bent downwards, impedance increases; i.e. a 50-Ω match can be made with a down angle of 42° (Figure 2.93).

Or alternatively, it can use an impedance-matching element in the antenna (usually a coil).

Can also fold the quarter wave antenna to increase impedance by fourfold, which can get the impedance close to 75 Ω, or by using smaller diameter grounded elements you can bring it to 50 Ω.

Its polarity is in the direction of orientation. It is nearly always used in vertical mode. It is widely used, simple and versatile.

2.14.3.4 5/8 λ Vertical Antenna

This vertical antenna is used in applications where an all-round radiation pattern is required, not just in horizontal; however, greater gain is afforded in the horizontal orientation.

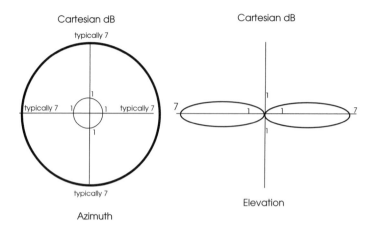

Figure 2.94   The 5/8 λ vertical antenna.

The peak gain is close to 4 dBd (relative to a dipole). This makes it particularly attractive for fleet mobile communications (Figure 2.94).

Matching is achieved by placing a small loading coil at the base to make it look electrically like 3/4 element, which has an impedence of 50 Ω.

It should be kept vertical and rigid to the ground plane to preserve the impedance.

2.14.3.5   Yagi Antenna

This is a more focused antenna. It comprises reflector elements which are about 5 % longer than the driven elements. The driven elements are arranged along a boom (Figure 2.95).

Parasitic elements (reflectors add inductance by coil or by lengthening element 5 %) reflect the main lobe.

Parasitic elements (director add capacitance by capacitor or shortening element 5 %) reinforce and add gain to the main lobe.

The gain is a function of frequency, not of number of elements. These days the antennas are computer-designed, using the physics of finite elements to optimize antenna.

Altering the element spacing is bound to affect impedance more than the polar pattern.

They can be tuned to 75 Ω, but more commonly they are tuned to 50 Ω.

If elements are spaced at intervals less than 0.2 λ, impedance falls rapidly away.

One can use folded dipole as driven element to increase the lowest frequency of operation.

It has a very narrow RF bandwidth and a good front-to-back ratio. (This is sometimes written as F/B ratio and is a measure of how much antenna gain goes out the front of the antenna compared with the back. The delta difference in the gains is called the F/B ratio in decibels.)

2.14.3.6   Log Periodic Antenna

Not really used in mainstream aeronautical communications other than for testing applications and in some HF applications (Figure 2.96).

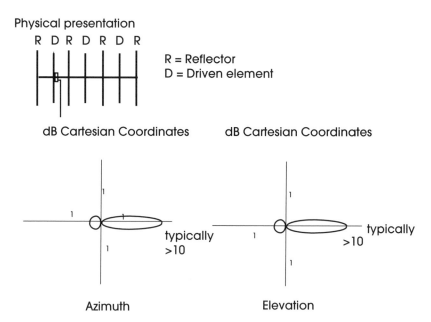

Figure 2.95 The Yagi antenna.

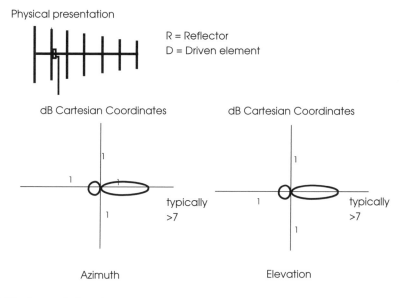

Figure 2.96 Log periodic antenna.

It has a very wide frequency bandwidth, typically 2:1 times its centre frequency to the 3-dB
  points.
Ideal for HF where frequency range considerable.
Still retains directivity and gain.
Good F/B ratio.

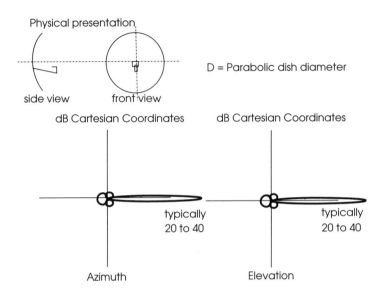

Figure 2.97 Parabolic dish antennas.

Generally less gain for size than its Yagi antenna equivalent.

Log periodic dipole array most common (elements diminish in size from back to front).

Element at the back of array where elements are largest is a λ/2 at back largest element (i.e. the cross metal objects).

Feed phasing alternated per element.

Stub matching or variable transformer for variable frequency use.

Typically 4–6 dB of gain over a bandwidth of 1:2, VSWR > 1.3:1.

2.14.3.7 Parabolic Dish Antennas

These antennas exibit the properties of light. The gain of such an antenna is given by the equation

$$G\,(\mathrm{dB}) = 10\,\log_{10} \frac{k\,(\pi D)^2}{\lambda^2} \tag{2.42}$$

where D is the antenna parabolic diameter and λ is the wavelength of operation (Figure 2.97). Most antenna efficiencies (k) are in the order of 55–65 %.

The antennas can be vertical or horizontally polarized (depending on the feed arrangement feeding the antenna).

There are many variations of this antenna, for example, the high-performance antenna (has a shroud that attenuates the off-centre beam radiation coming into the antenna) (Figure 2.98). Also, there are low-wind-loading or GridPack versions of this (Figure 2.99).

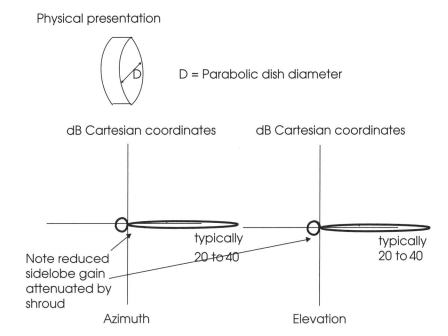

Figure 2.98   High-performance parabolic dish antennas.

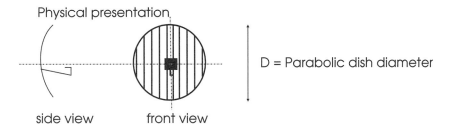

Figure 2.99   Low wind loading grid pack parabolic dish antenna.

2.15   The Link Budget

Consider a transmitter transmitting P_{Tx}. From Figure 2.100, the consequential receive power P_{Rx} can be found to be

$$P_{Rx} = P_{Tx} - L_{f1} + G_1 - f_{spl} + G_2 - L_{f2} \qquad (2.43)$$

Also,

$$\text{effective isotropic radiated power (EIRP)} = P_{Tx} - L_{f1} + G_1 \qquad (2.44)$$

And to recap from Section 2.4,

S (power flux density at input to receive antenna) $= \text{EIRP} - 10 \log(4\pi) - 20 \log d(\text{m})$

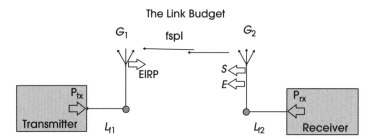

Figure 2.100   The link budget.

And the field strength at a point d from transmitter antenna

$$E(\mathrm{dB\mu V/m}) = \mathrm{EIRP(dBW)} - 20 \log d(\mathrm{km}) + 74.8$$

2.16   Intermodulation

Intermodulation is an extension to the principles of modulation; however, it is usually an unwanted process. It can be found to occur typically anywhere within a radio system transmitter, receiver or antenna system. Following are a few examples:

- where non-linear discontinuities exist such as in the non-linear part of a diode or transistor in a power amplifier when working in 'saturation';
- in a modulator that is not working properly (i.e. filtering characteristics are not properly filtering off wanted signal and rejecting unwanted signals.);
- where metallic elements or bad connections/junctions exist (e.g. with corroded feeder connections or bad grounding) in an antenna system;
- or from induction from adjacent high-power antennas into the antenna system;
- or through badly shielded receiver systems.

It is found that the odd harmonic products of the two or more input signals are most significant and harmful (Figure 2.101).

In particular, third-order harmonics are generally considered to be the most problematic ones.

2.16.1   Third-order, Unwanted Harmonics

Consider two input tones A and B. (Figure 2.102) The unwanted, third-order, two-station components lie at

$$2A - B \tag{2.45}$$

$$2B - A \tag{2.46}$$

Also usually out-of-band, third-order, two-station components lie at

$$2A + B$$

$$2B + A$$

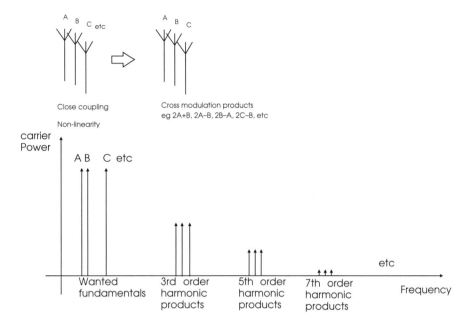

Figure 2.101   Intermodulation.

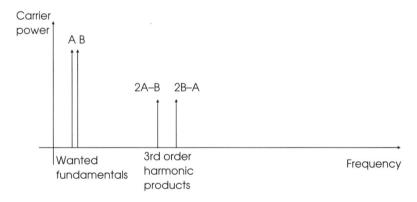

Figure 2.102   Two-station, third-order harmonic products.

Example 10 2-Station Example

Consider two frequencies at VHF: 131 MHz and 133 MHz.
   Then

Primary A	Primary B
131 MHz	133 MHz

	Unwanted intermods
2A − B	129 MHz
2B − A	135 MHz
2A + B	395 MHz
2B + A	397 MHz

In the case of the first two frequencies, both are in the VHF communication band (118–137 MHz) and therefore could be problematic to other users.

Extending this example to a three-station example: third-station, third-order, unwanted products. Assume the three primary frequencies of operation are A, B and C. Then the third-order intermodulation products will be

(double-product terms)

2A − B
2B − A
2A + B
2B + A
2A − C
2C − A
2A + C
2C + A
2B − C
2C − B
2B + C
2C + B

(triple-product terms)

2A + B − C
A + 2B − 2C
2A − B + 2C
A − 2B + 2C
2B + C − 2A
2C + B − 2A

Example 11

Consider three frequencies at VHF: 128 MHz, 131 MHz and 133 MHz.
  Then

Primary A	Primary B	Primary C
128 MHz	131 MHz	133 MHz

Unwanted intermods	
Double-product components	
2A − B	125 MHz
2B − A	134 MHz
2A + B	387 MHz
2B + A	390 MHz
2A − C	123 MHz
2C − A	138 MHz
2A + C	389 MHz
2C + A	394 MHz
2B − C	129 MHz
2C − B	135 MHz
2B + C	395 MHz
2C + B	397 MHz
Triple-product components	
2A + B − C	254 MHz
A + 2B − 2C	124 MHz
2A − B + 2C	391 MHz
A − 2B + 2C	132 MHz
2B + C − 2A	139 MHz
2C + B − 2A	141 MHz

See Figure 2.103.

Classification	TX – Frequencies	Examples
2nd order (2 stations)	A + B	131 + 121 = 252 MHz
	A − B	131 − 121 = 10 MHz
3rd order (2 stations)	2A + B	2 × 131 + 121 = 383 MHz
	2A − B	2 × 131 − 121 = 141 MHz
	A + 2B	131 + 2 × 121 = 373 MHz
	2B − A	2 × 121 − 131 = 111 MHz
3rd order (3 stations)	A + B − C	131 + 121 − 118 = 134 MHz
	A + C − B	131 + 118 − 121 = 128 MHz
	B + C − A	121 + 118 − 131 = 108 MHz
5th order (2 stations)	3B − 2A	3 × 121 − 2 × 131 = 101 MHz
	3A − 2B	3 × 131 − 2 × 121 = 151 MHz
5th order (3 stations)	2A + B − 2C	2 × 131 + 121 − 2 × 118 = 147 MHz
	A + 2B − 2C	131 + 2 × 121 − 2 × 118 = 137 MHz
	2A + C − 2B	2 × 131 + 118 − 2 × 121 = 138 MHz
	A + 2C − 2B	131 + 2 × 118 − 2 × 121 = 125 MHz
	2B + C − 2A	2 × 121 + 118 − 2 × 131 = 98 MHz
	2C + B − 2A	2 × 118 + 121 − 2 × 131 = 95 MHz
7th order (2 stations)	4A − 3B	4 × 131 − 3 × 121 = 161 MHz
	4B − 3A	4 × 121 − 3 × 131 = 91 MHz
7th order (3 stations)	3A + B − 3C	3 × 131 + 121 − 3 × 118 = 160 MHz
	A + 3B − 3C	131 + 3 × 121 − 3 × 118 = 38 MHz
	3B + C − 3A	3 × 121 + 118 − 3 × 131 = 88 MHz
	3C + B − 3A	3 × 118 + 121 − 3 × 131 = 79 MHz

Figure 2.103   Example intermods.

Thus in analysis, seven of these products fall into the VHF communications band (between 118 and 137 MHz) and could be problematic to other users.

2.16.2 Higher Order Harmonics

Odd harmonics seem to predominate in interference problems in radio systems. Generally the 5th harmonics have a lesser power spectrum than the 3rd harmonics and the 7th harmonics have an even lesser power spectrum than the 5th harmonics and so on and so forth. However, even 9th and 11th harmonic components can find their way into very sensitive receiver systems (for example, GPS receivers which work in CDMA below the noise floor) (Table 2.8).

Of course the same process can be extracted to three or more inputs, *ad infinitum*.

Table 2.8 Higher order harmonics.

5th order harmonic set
$\quad$ 4A $-$ B, 4B $-$ A, 4A $+$ B, 4B $+$ A
7th order harmonic set
$\quad$ 6A $-$ B, 6B $-$ A, 6A $+$ B, 6B $+$ A
N-th order harmonic set
$\quad$ (N $-$ 1)A $-$ B, (N $-$ 1)B $-$ A, (N $-$ 1)A $+$ B, (N $-$ 1)B $+$ A

2.17 Noise in a Communication System

2.17.1 Thermal Noise

Noise can be generated by several sources, including background noise (sometimes called thermal or white noise). This noise is that present in electronic circuitry by free electrons moving randomly in a conductor. It is a function of temperature; that is why it is called thermal noise.

$$N_{\text{thermal}} \text{ (sometimes called } N_0) = kTB \qquad (2.47)$$

where k is Boltzmann's constant, T is the temperature of operating in kelvin (K) and B is the bandwidth of the system under consideration in hertz (Hz).

2.17.2 Natural Noise

There are many other sources of noise, for example, natural sources, which include naturally occurring phenomena such as galactic noise and solar flare noise, weather activity (such as lightening discharges) and static discharges. These have to be considered individually in the bands that they affect.

2.17.3 Man-made Noise and Interference

Electrical noise can be generated from man-made sources such as industrial activity (e.g. industrial induction furnaces, microwave ovens, electrical motors and ignition systems, industrial electronics). Its attributes can usually be compared with those of white noise or an equivalent white noise source.

Noise caused by other man-made sources can directly interfere into a radio receiver. The characteristics can be such that they come straight in through the receiver bandwidth (e.g. electrical equipment with poor electromagnetic compatibility (EMC)). Also, other radio systems can produce noise or interference to the wanted radio system(s). These latter subjects will be discussed in much more detail in Chapter 12.

2.17.4 Sky Noise

Sky noise is the composite noise aggregation when looking from a receiver input up the antenna system. It includes the atmospheric noise components and the apparent amplification of these by a receiving antenna system.

2.18 Satellite Theory

This topic is dealt with almost last in this theory section of the book as it requires a prerequisite understanding of the free space path loss, link budgets, antenna systems and noise.

Satellite systems inherently deal with very low signal levels in space and on the earth's surface. Consequently very low noise amplification is required in the receiver portion with head-end electronics. High-gain amplifiers are generally deployed in the transmit portion of the system(s) and high-gain antennas are usually used (particularly on the ground where the physical constraints and economics are easier).

Link budgets are usually tight and critical and as previously discussed in the propagation section, (Figure 2.16) there are a number of natural windows in the spectrum for satellite operation. Where absorption losses are relatively low, in particularly, the L band around 1.4 GHz/1.6 GHz, C band around 4 GHz/6 GHz and at Ku band 12–14 GHz, all of these bands are exploited directly or indirectly for the use of aeronautical radio communications (Figure 2.104).

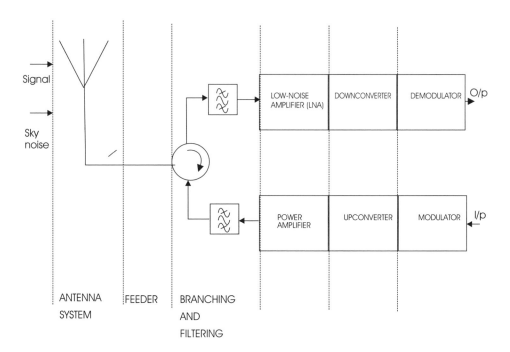

Figure 2.104 Satellite theory block diagram.

2.18.1 Extended Noise Equation

Referring back to the noise equation (2.47) and expanding this for unit bandwidth, an expression of Boltzmann's equation is obtained:

$$N_0 \text{ (unit noise in 1 Hz of bandwidth)} = kT_{\text{system}} \tag{2.48}$$

In logarithmic terms,

$$N_0 = -228.6\,\text{dBW} + 10 \log T_{\text{system}} \tag{2.49}$$

For receive signal power of C,

$$C/N_0 = C/kT_{\text{system}} \tag{2.50}$$

2.18.2 G/T

G/T is often called the 'figure of merit' of a receiver system and is a measure of a receiving system's ability or quality to demodulate a signal.

$$\frac{G}{T} = G\,(\text{dB}) - 10 \log T_{\text{system}} \tag{2.51}$$

where G is the receiving system antenna gain and

$$T_{\text{system}} = T_{\text{antenna}} + T_{\text{Rx}} \tag{2.52}$$

All noise temperatures: Antenna noise temperature includes all noise contributors plus sky noise up to the reference plane (usually the input of the low-noise amplifier (LNA)). The receiver noise temperature includes all noise sources from the reference plane to the output of the demodulator.

2.18.3 The Link Budget Equation

A revision to the equation developed in section 2.14 can be seen to be (by dividing throughout by N_0)

$$\frac{C}{N_0} = \text{EIRP} - \text{fspl} - \text{(other losses)} + \frac{G}{T_{\text{system}}}(\text{dB/K}) - k(\text{dBW}) \tag{2.53}$$

Other losses that may be included to the free space path loss could be

- polarization losses;
- pointing losses;
- off-contour loss;
- gaseous absorption losses;
- excess attenuation due to rainfall.

2.18.4   Noise Temperatures

It is convenient to process the satellite working system using the concept of noise temperatures. The relationship between noise figure and noise temperature is given by

$$\text{Noise figure (NF)} = 10 \log_{10}\left(1 + \frac{T_e}{290}\right) \tag{2.54}$$

where T_e is the effective temperature of the network element.

2.18.4.1   Receiver Side of the Reference Point

The compound noise temperature of a cascaded network can be calculated using the formula (Figure 2.105)

$$T_r = T_1 + T_2 + \frac{T_3}{G_1} + \cdots + \frac{T_n}{G_1 G_2 \ldots G_{n-1}} \tag{2.55}$$

For an attenuator such as a transmission line, the noise temperature of a passive attenuator is given by

$$T_e = T\,(L - 1) \tag{2.56}$$

where T is temperature and L is attenuation loss (in actuals, not in logs). This can then be substituted into the cascade formula (2.55) as necessary.

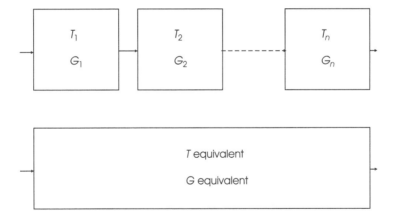

Figure 2.105   Noise temperature of cascaded network.

2.18.4.2   Antenna Side of the Reference Point

$$T_{\text{antenna}} = \frac{(L - 1)290 + T_{\text{system}}}{L} \tag{2.57}$$

where L is all system losses between antenna and reference point (Figure 2.106).

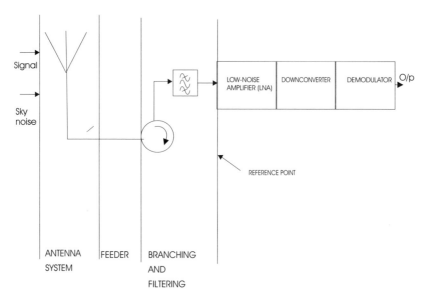

Figure 2.106 Receiver side and reference point.

Example 12

Satellite link budget
Consider the downlink from a satellite to an aircraft mobile earth station (MES) working in L band at 1.45 GHz. The EIRP from the satellite is 21 dBW, the range to the satellite is 25 573 NM, the aircraft MES terminal G/T is −13 dB/k (from the ICAO minimum specification for a high-gain antenna subsystem). Calculate the receiver C/N_0. Assume satellite pointing loss, off-contour loss, polarization loss, terminal pointing loss is 3 dB. (miles/km = 1.609)

Answer
Apply Equation (2.53)

$$\frac{C}{N_0} = \text{EIRP} - \text{fspl} - (\text{other losses}) + \frac{G}{T_{\text{system}}}(\text{dB/K}) - k(\text{dBW})$$

Or, in tabular addition/subtraction form,

EIRP	+ 21 dBw
fspl	−187.96
Losses + margin	−3.00
Terminal G/T	−13.00
Boltzmann's constant	−(−228.60)
C/N_0	45.82 dB

Figure 2.107 Example link budget to satellite mobile station.

Example 13

A video is being uplinked from the ground to an aircraft via a satellite. Consider the ground–satellite link working at 6 GHz. The terminal EIRP is 70 dBW. The required C/N_0 at the satellite is 102 dB. What G/T value must the satellite receiver system have? The satellite to earth station worst case distance is 37 000 km (note that not at the equator). Assume all other losses are 3 dB and a link margin of 3 dB is required.

Answer
Using Equation (2.53)

$$\frac{C}{N_0} = \text{EIRP} - \text{fspl} - (\text{other losses}) + \frac{G}{T_{\text{system}}}(\text{dB/K}) - k(\text{dBW})$$

and solving for G/T

$$\frac{G}{T} = \frac{C}{N_0} - \text{EIRP} + \text{fspl} + (\text{other losses}) + k(\text{dBW}) + \text{margin}$$

EIRP	−75 dBw
fspl	+199.34
Losses plus margin	+3.00
C/N$_0$	+102.00
Margin	−5.00
Boltzmann's constant	+(−228.60)
Terminal G/T	5.74 dB

Figure 2.108 Example cascaded network.

Example 14

Compute the system noise temperature for a three-stage receiving system. The first stage is an LNA, with noise figure of 1.1 and gain of 25 dB. There is a feeder between the stages of loss, with loss of 2.2 dB and the third stage has a noise figure of 6.

Answer
Compute each of the noise temperatures of the components individually, using Equations (2.54) and (2.56).

$$\text{NF} = 1.1 = 10 \log\left(1 + \frac{T_e}{290}\right)$$

Therefore $T_1 = 83.6$ K.

$$\text{NF} = 6 = 10 \log\left(1 + \frac{T_e}{290}\right)$$

Therefore $T_3 = 864.5$ K.
For the lossy transmission line,

$$L = \log^{-1}(2.2/10) = 1.66$$

Therefore $T_2 = (1.66 - 1) \times 290 = 191.3$

Similarly $G_1 = \log^{-1}(25/10) = 316.2$
and $G_2 = \log^{-1}(-2.2/10) = 0.603$
Plugging the values into the formula

$$T_r = T_1 + \frac{T_2}{G_1} + \frac{T_3}{G_1 G_2}$$

$$T_r = 83.6 + \frac{191.3}{316.2} + \frac{(864.5 \times 1.66)}{(316.2 \times 0.603)}$$

$$T_r = 91.73K$$

See Figure 2.108.

Example 15

Satellite
Consider the ground receiver of a satellite–ground link operating at 14.5 GHz. The antenna has a gain of 44 dBi. There is then a feeder with 0.4 dB of loss, then there is a feeder loss of 1 dB and a bandpass filter with 0.4-dB loss, random loss of 0.1 dB.

The LNA has a noise figure of 5 dB and a gain of 30 dB. It connects directly to a down converter/IF amplifier stage with a noise figure of 13 dB. Sky noise temperature is 18 K. What is the overall G/T of the system?

Answer
Net gain to reference plane

$$G_1 = 44 - 1 - 0.1 - 0.4 - 0.4 = 42.1\,\text{dB}$$

Sum of all losses to the reference plane

$$L_1 = 1 + 0.1 + 0.4 + 0.4 = 1.9$$

$$\text{Loss (actual)} = \log^{-1}\left(\frac{1.9}{10.0}\right) = 1.55$$

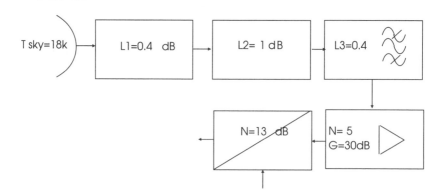

Figure 2.109 Example system block diagram.

Noise temperature of antenna subsystem to reference point

$$T_{\text{antenna}} = \frac{(1.55 - 1)290 + 18}{1.55} = 114.51 \text{ K}$$

Similarly

$$T_r = 627 + \frac{5496}{1000} = 632.5 \text{ K}$$

Overall

$$T_{\text{system}} = 114.51 + 632.5 = 747.05 \text{ K}$$

So

$$G/T = 42.1 - 10 \log(747.05)G/T = 13.6 \text{ dB/K}$$

2.19 Availability and Reliability

2.19.1 Definitions

Availability and reliability are different things. Availability provides information of how time is used and for what proportion of the time a system is operational or can be successfully used. Reliability conveys information about the failure-free interval. Both are normally quoted as percentages or decimals. For example,

$$\text{Availability} = \frac{\text{Uptime}}{\text{Uptime} + \text{downtime}} \tag{2.58}$$

2.19.2 The Reliability Bathtub Curve

It is well understood in the manufacturing industries, the aeronautical sector being no exception, that when equipment is manufactured, it is built to a design specification that has been written to optimize the quality of a product for a given price and attempts are made to ensure that it has a minimal failure rate and will last for a certain life span.

When the products of a sample are analysed as to how well they performed over a given period of time, it is often found (if the sample is large enough) that products generally follow the bathtub reliability curve (Figure 2.110); i.e. in the first instance there can be quite a high failure rate as products are 'burnt in'. To visualize this let us consider a mechanical example such as a piston in an engine. It takes a while for the surface areas of the piston and its new rings to adjust to other surface areas of the engine, and a gentle abrasion takes place on the brittle rings with the help of the oil. If inspected a few months later, the rings and piston would be a lot smoother to the touch. In the first few months or up to the first 10 000 km, motorists are encouraged not to over rev the engine and to take it easy whilst the motor settles down. Historically it is relatively likely for a machine to fail in its infancy due to an engine hotspot or in this 'burn in' process. Avionics and aeronautical radios are no exception.

In the steady-state period, the product is considered to have settled down and if it has survived the onerous stage of burn in, it is likely to survive the less onerous steady-state phase where the equipment is generally still new and durable. At the end of the life of most equipment, failure is due to aging of components from the continual wear and temperature cycling. This phase can be called 'wear out' or 'burn out'.

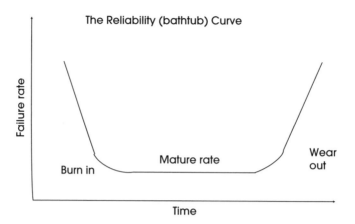

Figure 2.110   The reliability bathtub curve.

2.19.3   Some Reliability Concepts

Building on the idea of taking samples of an established product, one can build up the bathtub curve associated with each product. In addition, there are a number of statistical measures or concepts that form a basis for reliability engineering:

Mean time between failures (MTBF): This is usually measured in hours, and for a discrete component or system, it is the average lifetime of that component/system before complete failure.

This figure comes from the manufacturer and can be extrapolated from aging testing and also from feedback statistics as the product becomes mature. Obviously a fairly large sample of the product is needed to get something meaningful and representative. It is always interesting how a manufacturer is able to release statistics of the equipment sometimes before it has even gone to market. This should of course be questioned.

Mean time to repair (MTTR): This is the average time taken between a system 'failing' – till the fault is repaired – and the system being restored. It is measured in hours. With hardware, this normally involves a modular swap out or equipment change, which is a function of where equipment is deployed and where spares are held, getting them to site, etc. Usually it is of the order of a few hours. For the aeronautical case, a useful figure here can be the average flight time for the aircraft involved for a single sector. With software, normally it will be a system reboot so is of the order of minutes.

Taking both of these concepts into account, it is possible to predict the equipment availability.

$$\text{Availability } (A) = \frac{\text{MTBF}}{\text{MTBF} + \text{MTTR}} \qquad (2.59)$$

$$\text{Unavailability } (U) = 1 - \text{Availability } (A) \qquad (2.60)$$

Disaster recovery: This concept is the ability for a system or function to prevail after a major catastrophic event. In the ultimate reliability scenarios, complete duplication of all critical functions is carried out (e.g. control rooms, control towers).

2.19.4 Overall Availability of a Multicomponent System

This is straightforward mathematics.

2.19.4.1 *Serial Chain* (Figure 2.111)

This is where failure of any one component will render overall system unavailable. All elements are unduplicated and therefore unavailabilities are accumulative.

$$\text{So overall availability of A system} = 1 - (U_1 + U_2 + \cdots + U_m) \qquad (2.61)$$

Example

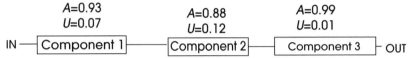

Figure 2.111   Reliability serial chain.

Total unavailability $= 0.2$, therefore total availability $= 0.8$.

2.19.4.2 *Parallel Chain* (Figure 2.112)

This is where elements of a system are duplicated and standby module takes over in the event of component failure. In this instance unavailabilities are multiplied together.

$$\text{So overall availability of A system} = 1 - (U_1 \times U_2 \times \cdots \times U_m) \qquad (2.62)$$

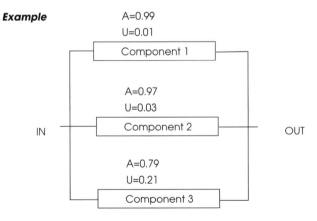

Figure 2.112   Reliability parallel chain.

$$\text{So overall unavailability} = 0.01 \times 0.03 \times 0.21$$

$$= 0.000063 \text{ or } 0.0063\%$$

$$\text{Availability} = 0.9999937 \text{ or } 99.9937\%$$

Serial chains and parallel chains can be considered to be analogous to resistors in series and parallel.

2.19.4.3   The Reliability Block Diagram

This is when you lay each of the components out in a diagram with each of the reliabilities, showing which elements are serial functions and which are parallel functions. From this an overall reliability can be computed.

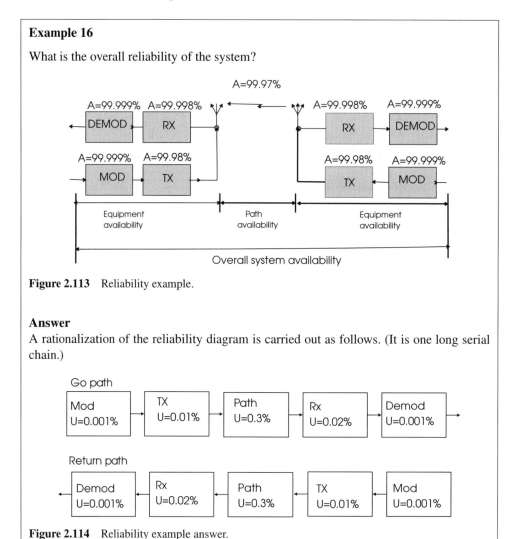

Example 16

What is the overall reliability of the system?

Figure 2.113   Reliability example.

Answer
A rationalization of the reliability diagram is carried out as follows. (It is one long serial chain.)

Figure 2.114   Reliability example answer.

$$\text{Total unavailability} = (0.001 + 0.01 + 0.03 + 0.002 + 0.001) \times 2^*$$

$$\text{Availability} = 1 - \text{Unavailability}$$

$$= 1 - (0.088)$$

$$= 0.956 \text{ or } 95.6\%$$

*Note Path unavailability needs counting in both directions.

Example 17

An airborne transceiver has an MTBF of 15 years. The corresponding ground transceiver has an MTBF of 20 years. The power supply for the airborne case has an MTBF of 35 years and for the ground case has an MTBF of 25 years.

For the airborne case, assume the average flight length in Europe is 2 hours. With the ground system, assume the base station site is 2 hours from the main communications depot holding spares, and the alarm being raised to mobilize the technicians averages 1/2 an hour.

If the path availability is 99.98 %, what is the overall system availability, ignoring man–machine interface (i.e. assume their availability is 100 %)? How would this change by duplicating the transmitter or receiver portions of this network? What would be the weak point in the system?

Answer

MTTR for airborne case can be taken as 1 hour (i.e. mid flight will be average) and assume spares are operational as it lands and docks.

Availability of airborne transceiver $= (15 \times 365 \times 24)/((15 \times 365 \times 24) + 1) = 0.999992$.

Availability of airborne power supply $= (35 \times 365 \times 24)/((35 \times 365 \times 24) + 1) = 0.999997$.

Availability of path $= 0.9998$.

Availability of receiver $= (20 \times 365 \times 24)/((20 \times 365 \times 24) + 2.5) = 0.999986$.

Availability of receiver power supply $= (25 \times 365 \times 24)/((25 \times 365 \times 24) + 2.5) = 0.999989$

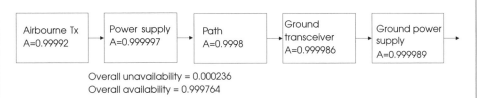

Overall unavailability = 0.000236
Overall availability = 0.999764

Figure 2.115   Reliability chain diagram.

Answer

The path availability dominates the overall system availability.

(Overall availability tends to 0.9998 or 99.98 % (i.e. that of the path). This is the weak chain in the reliability diagram.)

By duplicating equipment it would thus have a minimal impact on overall system availability which would remain at 99.98 %.

Further Reading

1. The Institution of Engineering and Technology, Savoy Place, London, see www.iee.org.uk
2. National Physics Laboratory, Teddington, Middlesex, UK, see www.npl.co.uk
3. International Telecommunications Union, Geneva, Switzerland, see www.itu.int
4. International Civil Aviation Organization (ICAO), Standards and Recommended Practices (SARPs) Annex 10. www.icao.int
5. ITU propagation recommendations, see www.iti.int

3 VHF Communication

Summary

This chapter looks at the band used for mainstream civil aeronautical communications in the very high frequency band between 118 and 137 MHz. It starts with the history and basics of the double side band amplitude modulated (DSB-AM) 25 kHz legacy system as it exists today. It looks at the radio system level, channelization plans and the evolution of channel splitting. It also builds on the principles of coverage, as drawn from the theory in Chapter 2, and principles of providing extended coverage and how this is done at a system level. It then looks to the interim future of 8.33 kHz voice working and VHF datalinks (VDLs). It does not discuss in detail any of the minor VHF and UHF bands above 118–137 MHz which are generally using the same technology for military or airport local area communications. These bands tend to be particular to the countries using them but the principles described in this chapter can generally be interpolated to these other bands.

3.1 History

3.1.1 The Legacy Pre-1947

Back in the early days of aviation, before radio was readily available, pilots used to dip a wing to signal to the control tower that they were coming into land on the next sweep past. This method was obviously open to various misinterpretations and the number of messages was limited to one – thus a requirement for better communication. The first recorded successful air–ground and ground–air radio transmissions were performed in 1917. However, the wing-tip-dipping still continued and it was not until the late 1920s early 1930s when it was to become technically and commercially feasible to install radio on-board an aircraft on a broader scale (see the foundation of ARINC[1] circa 1929); and it was not until the 1940s that radio became generally available for use on all aircraft. Moreover, it was proprietary, unstructured and by today's standards unreliable. It initially used the high frequency (HF) band between 3 and 30 MHz or even lower frequencies, which were prone to static and atmospheric noise.

In the 1940s the requirement for a reliable medium for communicating between aircraft and the ground was growing. At the same time the reliability of radio was increasing and the VHF band leant itself well (due to its relative immunity to atmospheric noise and reliable

Aeronautical Radio Communication Systems and Networks   D. Stacey
© 2008 John Wiley & Sons, Ltd

communications abilities for relatively low transmit power (typically 10–20 watts) and relatively high ranges (to the horizon) a number of things were to change.

At the end of the Second World War the invention of the jet engine opened a whole new opportunity to the passenger aviation era and enabled it to revolutionize the way of long distance travel. Then with the formation of International Civil Aviation Organization[2] (4th April 1947) it was recognized that a more structured approach to aeronautical communications was required. In parallel to the formation of ICAO and consequential to its new influence, the VHF band 118–132 MHz was set aside by the World Radio Conference in Atlantic City (1947) with the advent of the Aeronautical Mobile (Route) Service (AM(R)S)[3] for use in this band. The AM(R)S is essentially the same in principle today. By definition, this is for the use of 'a service reserved for communication relating to the safety and regularity of flight, primarily along national or international civil air routes'.

It is worth mentioning that the AM(OR)S or Aeronautical Mobile (Off-Route) Service was established at the same time and 'intended for communications, including those relating to flight coordination, primarily outside national and international civil air routes'; i.e. it was intended for the use of military communications. It is discussed further in Chapter 4.

The new AM(R)S system was planned to use 200 kHz channel spacing; thus 70 channels could be accommodated over the whole band. The modulation scheme to be employed was DSB-AM – mainly due to its simplicity and resilience in the environment it was to be operated in. It should also not be forgotten that back in the 1940s this was still cutting edge technology, and simplicity, reliability and costs were the key driving forces.

It was operated in dual-simplex mode (sometimes called half duplex). That is, only one transmitter could be operated on the channel at any one time and initially the word 'over' was used to signal to other users that they were shutting down their transmitter and it was available for the return path or anyone else (Figure 3.1). Also it was operated in a non-trunked fashion, i.e. when the air traffic control (ATC) or mobile transmitted, it was a 'broadcast' to all receivers in range. This had the added operational advantage that everyone listening knew what was going on.

3.1.2   1947 to Present, Channelization and Band Splitting

As time went on, the market demand grew and in some areas of the world the 70 channels were fast being taken up and it was considered that plans had to be made to accommodate the eventuality of radio congestion and even saturation. At the same time the radio technology was continually improving. There were two criteria that were ultimately defining the bandwidth required by each of the radio frequency (RF) channels.

Carrier frequency accuracy. The carrier frequency accuracy that could be achieved by both ground and airborne transceiver equipment. This was becoming better all the time with technological improvement making it feasible to contain the modulated signal in a narrower band than the 200 kHz originally allocated.

The actual bandwidth occupied by the DSB-AM spectrum. It should be noted that a typical unmodulated voice signal shows a spectral density with most of its power components above 200 Hz and below 4 kHz (Figure 3.2).

Arguably, the spectrum differs slightly with male (more low-frequency components) and female (more higher frequency components) but in general for both sexes, most of the power spectrum density is retained under the 4 kHz portion of the graph.

There are some minor components as high as 20 kHz, but between 5 and 20 kHz there is negligible power, and even if there was, the typical human ear loses sensitivity to these tones. So (today) it is possible to accurately re-create a voice signal by demodulating a signal with

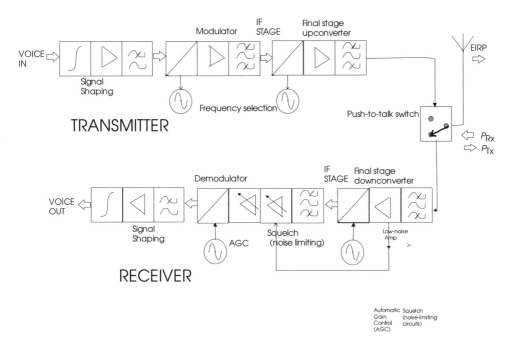

Figure 3.1 DSB-AM radio half-duplex operation.

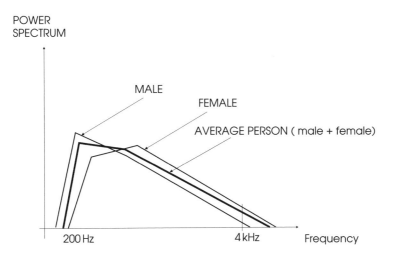

Figure 3.2 Typical voice spectral density.

the carrier plus all the components band limited ±4 kHz. (In comparison, for the music purists even CD quality is only ±8 kHz.)

If a DSB-AM signal is created by modulating a carrier with a voice-limited signal to ±4 kHz, the modulated spectra will similarly be limited to ±4 kHz about the carrier.

3.1.2.1   Channel Splitting

Taking both the factors, *channel frequency accuracy* and *the actual bandwidth occupied by the DSB-AM spectrum*, into account it is possible to reduce the channel spacing and hence to increase the amount of channels available in the given spectrum. This is exactly what happened, not just once but a number of times as the technology improved and the demand on channels continued as growth in civil aviation continued.

In particular in the 1950s, –100 kHz channel spacing was first introduced, which doubled the capacity to 140 available channels. In addition WRC 1959 further extended the band allocated for AM(R)S to 118–136 MHz; this meant 180 channels at 100 kHz were achievable.

In the 1960s, this methodology was extended further, with 50 kHz channel spacing now easily achievable, and this doubled the capacity again to 360 channels at the rate of 50 kHz. As time went on, the market grew further and the technology improved. . . .

In 1972, –25 kHz channel spacing was introduced and this doubled the capacity again to a theoretical 720 channels. This alone was not enough to curb demand and in 1979, WRC extended the AM(R)S allocation in the VHF band further to 117.975–137.000 MHz, which is where it is today with a theoretical 760 channels at the rate of 25 kHz achievable.

A further reason for the channel splitting approach was backward compatibility between old and new radio systems.

3.1.3   Today and 8.33 kHz Channelization

In 1996 again, driven by an increase in air traffic and consequently demand on VHF channels, a further channel split to 8.33 kHz was proposed in Europe only. This gives a theoretical 2280 channels achievable. In the United States, there was reluctance to go with this channelization as it was believed a superior digital system (namely VHF datalink 3 or VDL3) would supersede the legacy VHF DSB-AM analogue system. The choice of 8.33 kHz was chosen as it was the minimum practical size to support DSB-AM modulation (if the ±4 kHz voice limiting is applied as discussed above, a modulated channelization of 8 kHz can be obtained), and it theoretically provided a threefold increase in voice channel capacity. Further band limiting voice below 4 kHz would mean degradation in voice quality. So despite some interests saying it could be further channel split past 8.33 kHz, this is likely to be the last realization of channel splitting and the end of the road for the DSB-AM technology.

Also, it should be pointed out that realizing 2280 channels is 'theoretical' – this is an oversimplification. The practical reality is that this number cannot be reached for a number of reasons, such as sterilization of channels adjacent to 'protected' or high-priority services (such as airfield and sector ATC frequencies), the overlay of 25 kHz VDL channels (making these out of bounds for conversion to 8.33 kHz working), a reluctance for some aircraft to move to 8.33 kHz channelization which has left a legacy requirement to retain some 25 kHz channels and also some of the practical elements of coverage and range of services (for example, CLIMAX that will be discussed in much greater detail later). There are also constraints placed on new services from old allocations and service volumes.

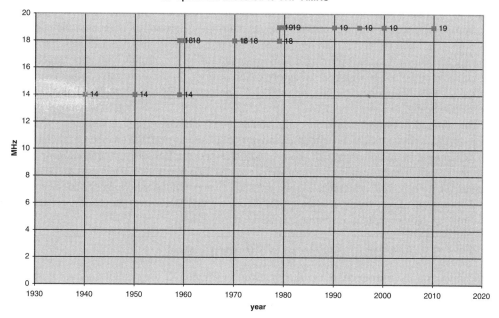

Figure 3.3 Channelization evolution.

The sad reality of all this is that in practical terms not even two 8.33-kHz channels' assignment can be created or 'converted' for the one legacy 25-kHz channel assignment, and even this is proving hard as the congestion problem within core Europe mounts up and compounds. The ICAO Europe Frequency Management Group (FMG)[4] that is responsible for implementing the transition from 25 kHz working to 8.33 kHz working has regular planning meetings to do these conversions, but with these constraints it is showing diminishing returns. A radical rethink is required.

Also it is unfortunate to see that aviation interests have fragmented into different regional policies, with Europe backing the 8.33 kHz deployment and North America backing the seemingly oppositional VDL3 solution (using 25 kHz channelization) rather than one international system deployed by all. This complicates equipage issues and future options, as will be seen later.

3.1.4 Into the Future (Circa 2006 Plus)

The legacy analogue frequency division multiplexed AM(R)S band 117.875–137.000 MHz still stands today and the current thinking within the industry is that it will last in its 25-kHz and 8.33-kHz co-habitation form for at least 20 more years. In some of the less congested areas of the world, 50 kHz and even 100 kHz channelization still prevail (for example, Africa and New Zealand). Within the AM(R)S there are provisions for VDL services, which will be discussed shortly. The band splitting limit has, for practical purposes, been reached at 8.33 kHz and the next step to fix increased demand will have to be a radical new re-engineering of the AM(R)S and a new system. The options for this are discussed in detail in Chapter 8.

8.33 kHz implementation status. Despite paving the way for 8.33 kHz radio implementation back in the mid-nineties, as with many aviation project cycles, the implementation of 8.33 kHz in Europe is still very much in its early stages. Currently Europe has mandated the carriage of 8.33 kHz radio by all civilian and military aircraft above flight level 195 (19 500 feet). Except for some minor exceptions for some state aircraft or for aircraft with a low amount of anticipated air hours through the mandatory coverage area, for these aircraft continual support to the 25 kHz channelization will be required. The current 8.33-kHz deployment region is described in Figure 3.4.

Decisions as to further reduce the ceiling on 8.33 kHz expansion (called vertical expansion) or to further enlarge the area of effect (called horizontal expansion) have yet to be taken by the ICAO Europe in conjunction with the ECAC[5] states and the European commission in 2007, but the benefits would seemingly diminish against the significant increase in cost in these next stages. There is a great reluctance by general and private aviation who have to date been able to defer the equipage. Today, it is really just the airline and to a lesser extent some private jets and military/government aircraft that have 8.33 Hz installed.

3.2   DSB-AM Transceiver at a System Level

Recapping on the introduction and theory given in Chapter 2, the system diagram is shown below. This diagram is applicable for 200, 100, 50, 25 and 8.33 kHz channel working. All use DSB-AM (Figures 3.5 and 3.6).

It should be noted that modern receiver design may bypass the two-stage intermediate frequency process (as is defined in ICAO) and do it in a single voice frequency to radio frequency stage, or in some of the latest cutting-edge transceivers described later, the synthesis can even be carried out in a software on a modern processor using a fast sample period (multiple of the basic carrier rate) instead of the traditional crystal filter resonant circuits with harmonic generation synthesis.

The difference between the channelization being used will be reflected in the band-limiting filter at the end of the signal-shaping stage. Here the upper bandwidth cut-off of this filter must by definition be a maximum of half of the channel spacing. So at 25 kHz, it must be 12.5 kHz and for 8.33 kHz it must be 4.165 kHz in its ideal form (4 kHz is used in practice). Similarly the corresponding receiver should have a matching low-pass filter as per above in its signal-shaping circuitry dependent on the channelization being used. It will also be reflected in the thumbwheel dial selector used in the cockpit, which selects the channel frequency. On the ground the base station is usually fixed frequency working.

As previously pointed out, the majority of the voice power spectrum is contained below 4 kHz, so for the 8.33 kHz system the voice signal can be reconstructed with most of the information being retained and just a minor degradation in quality to what was sent. Operationally the degradation with 8.33 kHz is almost negligible to all but the most trained ear and as such it has been operationally accepted by all the regions shown in the map in Figure 3.4.

Also because of this voice power spectrum, it means that an 8.33 kHz radio is capable of receiving signals sent by 200, 100, 50 and 25 kHz working. Indeed this 'backward compatibility' of DSB-AM has been a key feature in its evolution. Similarly a 25 kHz radio would be able to receive an 8.33-kHz channel signal, provided it was tuned to the same carrier frequency.

3.2.1   System Design Features of AM(R)S DSB-AM System

For the radio system designer or planner, a couple of system features should be pointed out.

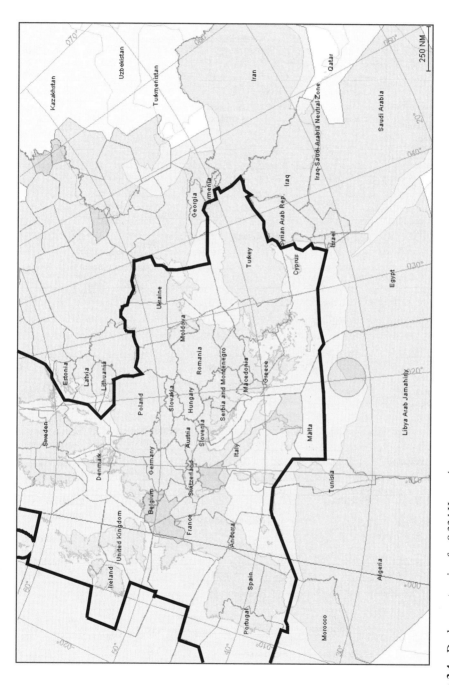

Figure 3.4 Deployment region for 8.33 kHz service.

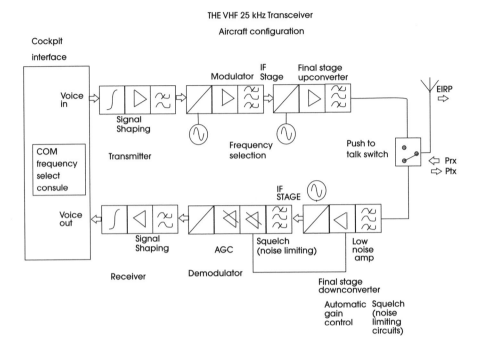

Figure 3.5   DSB-AM aircraft transceiver.

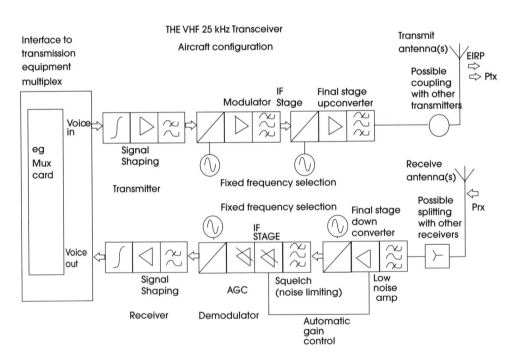

Figure 3.6   Corresponding DSB-AM ground transceiver base station.

3.2.1.1   *Availability and Reliability*

It is hard to find an international standard for system reliability or availability for a link from aircraft to ground, ground to air, or air to air. This is usually one of the most fundamental design questions or specifications to the radio system designer. If the ICAO Standards and Recommended Practices (SARPs) Annex 10 (Chapter 2)[6] is consulted, it specifies a specific field strength at the receiver antenna 'on a high number of occasions'. It would seem that system reliability and availability is a national issue and this is where it is usually documented. On consultation with a number of national administrations, different countries seem to design for different availability criteria, which have been seen to be as diverse as from 99.9 to 99.999 % in each direction.

3.2.1.2   *RF Unbalance*

In addition, the requirement on the airborne transceiver equipment is less onerous than on the ground equipment. This is partly a legacy aspect, but also makes sense to continue this, as practical realization in the air is harder than on the ground. Also the ground portion is of fixed frequency where the airborne unit needs to be tunable across the band. More specifically these are the specifications on each equipment on 'a high number of occasions' (again taken from ICAO SARP's Chapter 2) (Tables 3.1–3.4).

There are a couple of legacy elements to be considered from this specification.

Firstly these specifications were written when the technology was still restricted and it was easier to build higher power transmitters and more sensitive receivers that were ground based; also the ground-based antenna pattern can be focused to give an upward beam maximized for horizon as that is where the link budget will be most limited.

With the aircraft transceiver system, power, size, weight and allowable heat dissipation were and still are a luxurious commodity. With the antenna system, it is required to use the closest thing to an omnidirectional antenna to allow for all pitches and rolls of the aircraft. Hence a more lenient set of parameters to the airborne side of the system is required. Additionally the more lenient requirements for extended coverage are discussed later.

3.2.1.3   *System Specification*

Other generic system design specifications as laid out in ICAO SARPs are as follows (Table 3.5).

The adjacent channel rejection ultimately defines the design of the voice-limiting filter and the filter after the RF stage.

3.3   Dimensioning a Mobile Communications System–The Three Cs

Something gleaned from experience in other sectors of the radio engineering industry and in particular the public mobile communications community is a concept called the three Cs. These stand for coverage, capacity and 'cwality' (OK! the last one is mis-spelled and should be quality but it is easier to remember in the acronym as the three Cs). In summary these are the three determining criteria for any mobile communication system. This can be applied to the aeronautical mobile scenario. In the following, these aspects are considered in detail.

Table 3.1   Transmitter ground specification.

- Frequency stability ±0.005 % (well within Doppler effects as well)
- Superheterodyne synthesis (2 stage)
- Designed to give field strength of 75 µV/m (−109 dBW/m^2) on a 'high number of occasions', with modulation depth of greater than 85 %

Table 3.2   Receiver ground specification.

- Field strength of 20 µV/m (−120 dBW/m^2), to provide a wanted/unwanted output voice signal of >15 dB (with depth of modulation 50 %) on a 'high percentage of occasions'
- Frequency stability ±0.005 % (well within Doppler effects as well)
- Adjacent channel rejection >60 dB
- Next assignable frequency is two channels away
- Squelch lift (receiver muting) 14 µV/m

Table 3.3   Transmitter air specification.

- Exactly the same as ground system with following differences:
- Less stringent frequency stability for 25 kHz, 0.003 %
- Designed to give field strength of 20 µV/m (−120 dBW/m^2) on a 'high percentage of occasions at appropriate range and altitude'

Table 3.4   Receiver air specification.

- Exactly the same as ground system with following differences:
- In the sensitivity allow for feeder mismatch and loss of antenna polar diagram variation
- With field strength of 75 µV/m (−109 dBW/m^2), to provide a wanted/unwanted output voice signal of >15 dB (with depth of modulation 50 %) on a 'high percentage of occasions'
- For extended VHF facilities a sensitivity of 30 µV/m should be assumed
- Squelch lift (receiver muting) 14 µV/m

Table 3.5   System specification.

- Adjacent channel rejection
  50 dB at ±25 kHz
  40 dB at ±17 kHz
- Emission designation A3E, service defined as AM(R)S (per ITU definition)
- Polarization vertical

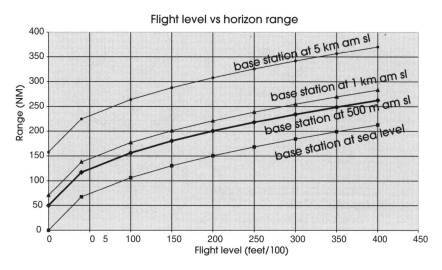

Figure 3.7   Relationship between station height and range.

3.3.1 Coverage

This is the service area of a generic mobile system. Generically it could be split into typical rural, sub-urban or urban coverage areas, or in aeronautical equivalent terms as long distance en route coverage, terminal manoeuvring area (TMA) coverage or local airport coverage.

In the initial design of a system, network designers tend to go for quick coverage or saturation of the area concerned. More mature networks go on to optimize this for the services required and to minimize self-interference to other parts of the network.

Recapping from the Theory section (Chapter 2), there is a relationship between height of the transmitting and receiving antennas (or flight level in the case of the airborne transceiver) and line-of-sight (LOS) distance or radius of coverage (Figure 3.7).

In general terms, the LOS distance can be considered as the limiting factor for long-distance VHF communications. (Over the horizon, the obstructed radio path sees dramatic increase in the attenuation; thus the effective usable coverage boundary is usually close to the horizon). This is applicable for the en route phases of flight and the upper airspace.

However, this LOS model is not appropriate for non-LOS communications that can occur for low-altitude flying (particularly in mountainous terrain) such as in the TMA or in the vicinity of the airport and on the ground. For these occasions beams can be refracted or reflected and the complex two-ray model such as proposed by Hata and Okamura[7] is more appropriate. As a yardstick, obstructions and knife-edge or smooth diffraction can typically increase the natural free space path loss between two LOS points by tens of decibels.

This sets some broad guidelines considering the system design.

- For upper airspace, a LOS link can be assumed to be ultimately limited by horizon and link budget (some further consideration should be given in mountainous areas or overseas) as described in the Theory section. A conservative *k* factor is that of 2/3, allowing for refraction

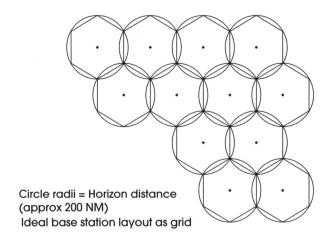

Figure 3.8 A topology for blanket coverage of upper airspace.

towards the earth. It generally gives a good conservative estimate to which horizon communications can be established and maintained for a very-high availability factor (>99.9 %)
(Figure 3.8).

• For ground links and low-airspace links, careful attention should be given to terrain, building
clutter and any other likely obstructions, leading to links usually not being direct LOS
links. Link budget design tools fitted with terrain and building geographical/topographical
information should be used if available (Figure 3.9).

In both instances, nothing can substitute for practical testing after a design has been decided
to verify critical or marginal areas (Table 3.6).

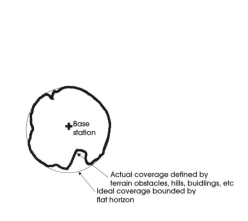

Figure 3.9 A typical coverage topology for lower airspace (for the same region).

Table 3.6 Summary ICAO Annex 10 specification for coverage.

- For en route, up to 200 NM (or 350 km) basically LOS which is a function of antenna height at both ends of the path (as already discussed, see ICAO Annex 10, attachment B to Part 1, Clause 4.4.3).
- For some ground communications, the coverage can be limited by path obstructions and reflections (for example, at a large airport) and a number of solutions can be used such as multiple base stations (and voting networks), multiple channels.
- ICAO defined the field strength required to be of 75 µV/m (–109 dBW/m^2) for ground to air and 20 µV/m (–120 dBW/m^2) for air to ground on a 'high percentage of occasions' for standard working. (See also specification for extended coverage where these specifications are reduced.)

3.3.1.1 *Voting Networks and Extended Coverage*

Looking back at the problem with low altitude where LOS conditions may not exist, there are two techniques widely deployed to overcome this problem to some extent without having the full cost and operational implications of straight cell splitting and full equipment duplication and operation.

3.3.1.1.1 Frequency Coupling

Imagine a number of busy operational sectors, each with their own team of air traffic controllers. Now imagine the situation when the out-of-hours traffic is significantly down on peak instantaneous aircraft counts (PIAC) and only one operator is required to cover a number of sectors. The two frequencies can be 'coupled' such that all ATC information is transmitted out on both frequency channels. On the receive side the controllers' console selects the voice stream of whichever is picked up or the stronger of the two signals by using a voting circuit (Figure 3.10).

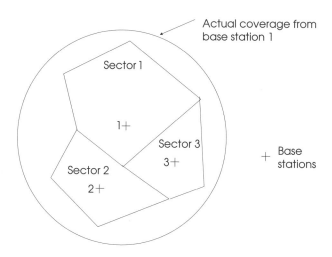

Figure 3.10 Frequency coupling.

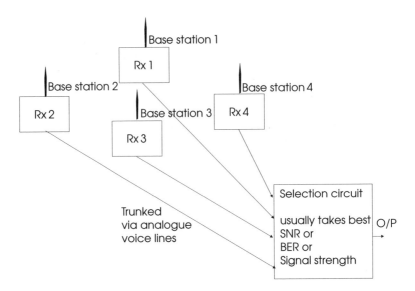

Figure 3.11   Receiver voting.

Voting can be selected by signal strength (the usual method) or from bit error rate (BER) analysis, which is usually more relevant to digital systems. The concept in operational terms is called 'collapsing sectors' and for analogy is also often used frequently in other industries (the oil industry, electricity industry, etc.) In some cases where the frequency being operated in one cell gives adequate coverage into the next cell, the second cell frequency can be altogether switched off (Figure 3.11).

3.3.1.1.2   Extended Coverage for a Cell Using Offset Carrier Techniques (CLIMAX)

In some instances the topology of a sector may be such that it does not easily lend itself to coverage by one base station. Consider a situation where a particular sector (shown as a pentagon in ground area) requires four separate transmitting base stations to give it saturated coverage (or the availability defined by the national regulations) (Figure 3.12).

For a controller to have to use multiple radio channels to cover this area would be cumbersome in operational terms.

In the 25 kHz environment with the progress of technology, a solution is available using multiple base stations to give blanket coverage. It exploits the characteristics of the receive filter and the legacy situation previously discussed. It also builds on the capture effect (sometimes called the FM capture effect although it is equally attributable to AM systems) that once a receiver demodulator locks onto one carrier, it will hold it in favour of a secondary (that is initially weaker but can become stronger see hysteresis loop in theory section) carrier.

To minimize interference and cancelling, the voice signal is simultaneously sent out from a number of base stations but with the carrier offset relative to each other (Figure 3.13).

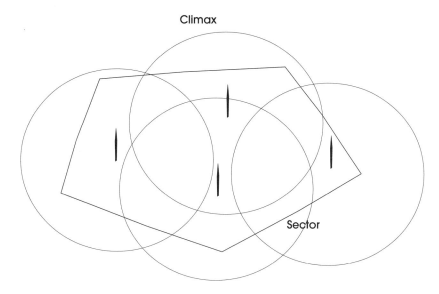

Figure 3.12   Extended coverage (CLIMAX) operation.

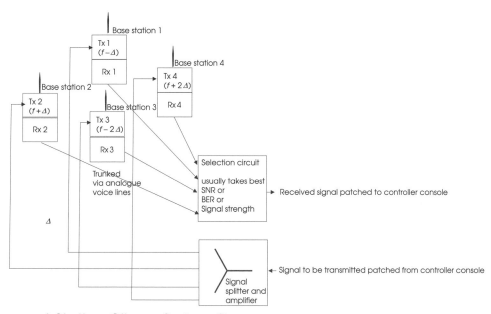

Figure 3.13   CLIMAX system diagram.

Table 3.7   CLIMAX configurations.

Number of base stations in a sector	Frequency offset
1	0
2	$f_1 = f_c + 5$ kHz
	$f_2 = f_c - 5$ kHz
3	$f_1 = f_c + 7.5$ kHz
	$f_2 = f_c - 7.5$ kHz
	$f_3 = f_c$ kHz
4	$f_1 = f_c + 7.5$ kHz
	$f_2 = f_c - 7.5$ kHz
	$f_3 = f_c + 2.5$ kHz
	$f_4 = f_c - 2.5$ kHz

There can be between two and four offset base stations or 'CLIMAX' systems (Table 3.7). The offset is defined as described in Table 3.8.

Again on the receiver circuitry side, a simple voting system can be used to elect which base station is receiving the strongest signal and switch this into operation.

Table 3.8   ICAO capacity specification.

- There isn't one as yet.
- The legacy VHF network is simplex, non-trunking with up to a number of users per channel managed by a controller, etc.
- Busy sectors/airports may run multiple ATC/AOC channels – rural areas might just have one channel for all operations.
- Capacity is fast becoming the main issue with the current VHF AM(R)S system to congested core flight areas.

3.3.2 Capacity

Capacity can be the quantity of mobile calls being carried out on the total network or in an individual cell of the network. Or it can be defined as number of voice calls per area (erlangs/km^2) or the amount of throughput required per mobile per cell, etc. Usually networks are designed for the busiest hour or in aviation terms, the PIAC (the Peak Instantaneous Air Count within one operational sector). This is the number of aircraft in the sky within a given service volume at a given time.

Here the nomenclature difference in aeronautical radio terminology between a sector and a cell should be noted:

A *sector* is an operational term for a delineated volume of 'airspace,' across which air traffic control (ATC) makes responsible one controller.

A *cell* is a radio coverage volume where one base station can provide coverage as limited by the link budget equation (cells are usually circular but are not necessarily so if terrain delineated or using directional antennas).

Capacity is becoming a huge issue in the core area of Europe and the eastern seaboard of the United States, parts of the US Californian area and in parts of Japan.

A number of concepts are used to cater for high-density areas. To list a few:

1. *Channel splitting*. That is where an ATC centre splits its operational traffic over a number of different control channels. In less congested airspace over a rural area, one channel may be used for everything. Or channel splitting as described above (i.e. 25 kHz to 8.33 kHz) is another manifestation of this same concept.
2. *Self-interference control*. This can be a complex engineering exercise where interaction between areas using the same frequency is minimized. Ideally, this is reduced to zero; in practice, this can very rarely be totally eliminated. It should be noted that the ICAO has planning rules for defining co- and adjacent-channel separation. (Note that these criteria are different in different world regions; e.g. Europe has a more stringent set of rules compared to those used in the United States – another legacy aspect.)
3. *Channel pinching or 'cherry picking'*. This is a cellular concept, where channels normally deployed on cellular reuse pattern that would have normally been allocated in a rural area can be 'taken away' and deployed in capacity hotspots such as large airport TMAs.
4. *Protection*. Some channels are of higher priority than others; i.e. the ones used by ATC on approach and landing are usually of higher importance than those for airline operational communications (AOC). Hence adjacent channels to the former are blocked from use. In the case of the emergency frequency (121.5 MHz), two 25 kHz channels on either side are blocked.

To give the reader a feel for capacity, it should be noted that in Europe there are over 10 000 allocations made to different services in the AM(R)S band (117.975–137 MHz). This equates to each channel (assume 25 KHz) being allocated $\approx$ 13 times on average.

In radio hardware terms, a way of resolving capacity on-board an aircraft is by multiple radio transceivers working different channels (sometimes denoted comm1, comm2, etc.) On the ground, it can be by stacking multiple base station transceivers in an equipment room and coupling these onto the necessary transmit and receive antennas (Figure 3.14).

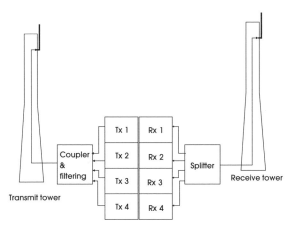

Figure 3.14 Multiple base station configuration.

Table 3.9   Quality specification.

- Availability and reliability of a channel is poorly defined in the ICAO SARPs, 'on a large number of occasions', ANSP's to interpret this as 99.9–99.999 each way. (this is path and equipment reliability.)
- The blocking probability for a network connection and once a connection is established, the chances of maintaining it is more appropriate to trunked communications. The aviation system is currently non-trunked in communication terms; the arrival stack effectively is a form of trunking and this is where the probability of blocking occurs not in the communication.
- The demodulation of voice for a signal-to-noise ratio (SNR) of 15 dB, 85 %.
- For data a BER of between 10^{-4} and 10^{-6} is usually specified.

3.3.3 Cwality (Quality)

Quality was partly touched on in the ICAO specifications previously. It can be many things:

- Availability and reliability of a channel.
- The blocking probability for a network connection and once a connection is established, the chances of maintaining it (probability of drop-out).
- The quality of the channels available either in signal-to-noise ratio (i.e. signal-to-interference ratio) or in future digital radio terms of BER.

Traditionally when rolling out, a network coverage is the first consideration. In busy parts, the capacity issues can quickly become priority and override. The quality is usually the third and final consideration but still importantly it needs to be defined for the services and applications being run (Table 3.9).

Example 1

1. To give complete upper airspace coverage for Europe, approximately how many base stations would be needed? (Assume Europe is a square 4000 km by 4000 km.)
2. What proportion of area cannot be reached by VHF with good reliability/availability?
3. How would you communicate with this area?

Answer
To oversimplify

- Consider the area of the hexagon $= 1.5\,R^2$;
- Consider a hexagon with 350-km radius;
- This has an area of $1.5 \times 350 \times 350 = 183,750\ \text{km}^2$;
- Area of Europe is approximately $16\,000\,000\ \text{km}^2$;
- If the hexagons are overlaid contiguously on the map of Europe
- Number of cells (hexagons) $= 16\,000\,000/183\,750 = 87$.

Approximately one-quarter is sea, most of which is within 200 NM of land. So some of this area can be covered by VHF or indeed probably most with some careful planning; for those parts that cannot be covered, this is covered by HF and satellite communications as discussed in the next chapter.

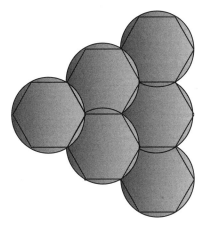

Figure 3.15 Contiguous coverage (circle/hexagon mapping).

Practical reality. Note: From legacy, the VHF coverage upper airspace cells for Europe are usually mapped to operational sectors and national borders. It is also interesting to note that Europe has between 5 and 10 times this number just for upper airspace. This is because of the non-contiguous boundaries between countries, the areas above sea and the legacy aspects of this. Also, for lower flight level services a reduced radius cellular pattern would be required. In addition, where there are capacity issues, more than one frequency is used per sector especially for vital services. Finally, the critical ATC channels are duplicated for reliability (quality purposes).

3.4 Regulatory and Licensing Aspects

3.4.1 The Three As

In contrast to the three Cs, we have the three As, which pertain to the regulatory and licensing aspects of the VHF band. They stand for the following.

3.4.1.1 Allocation

This is how the band 117.975–137.000 MHz is legally apportioned for use only by the AM(R)S on an international basis. This allocation was made and revised at World Radio Conferences (WRCs are administered by the International Telecommunication Union, Radio division (ITU-R); this organization carries UN status.) The ITU represents the general will of the state radio communication agencies as they strive for standardization and good compromise between radio interests. The outcome, the 'ITU radio regulations' is signed in by all its member states (approaching 400 currently) as national policy by the radio administrations in those countries. More detailed analysis of this can be found in the ITU radio regulations available for sale from the www.ITU.int website.

Table 3.10   Generic allotment of VHF channels for Europe (118–137 MHz).

ALLOTMENT OF 117.975-137 MHz				
FROM	TO	Channels (25 kHz)	Guard bands	ALLOTMENT
118	121, 4	136		International and National Mobile Services
119,7	119, 7	1		reserved for regional guard supplementary tower and approach services
121.5	121.5	1	+/–3 channels	Emergency Frequency
121,6	122	15		International and National Aerodrome Surface Communications
122	123, 05	42		National Aeronautical Services
122,1	122, 1	1		reserved for regional guard supplementary tower and approach services
122,5	122, 5	1		reserved for regional light aircraft
123.1	123.1	1	+/–1	International S&R, (ATIS allowed in guard bands)
123,45	123, 45	1		Air to Air communications for remote and Oceanic
123,5	123, 5	1		reserved for regional light aircraft
123,15	123, 6917	21		National Aeronautical Mobile Service
129,7	130, 8917	47		National Aeronautical Mobile Service
130,9	136, 875	239		International and National Aeronautical Mobile Service
131,4	131, 975	23		reserved for operational control communications
131,525	131, 725	8		reserved for ACARS datalink
136,8	136, 875	3		reserved for operational control communications, but no new assignments
136,9	136, 975	3		VDL
note also 136, 5-136, 975 inclusive not for the use of 8, 33				

3.4.1.2   Allotment

This is the second tier, and this is how the spectrum assignment is cut up and channelized. This is usually a collaboration between the ITU-R, radio administrations in each country and ICAO. The current allotment for this band generally adopted by most nations is described in Table 3.10.

3.4.1.3   Assignment

This is the third tier and the process of individually licensing transmitter equipment. This is done by the administration where the equipment is operated, or in the aviation case, the country where the aircraft is registered. Legally they are allowed to use a transmitter registered in one country in another country through the special provisions defined in the ITU-R radio regulations. Usually a register of the licenses is held by the country of licensing and the interested party.

3.4.1.4   Utilization Profile

This graph is a simplified representation of the information contained in the European ICAO Comm 2 tables (Figure 3.16).

Looking more closely into the 10 000 plus assignments in Europe, it is interesting to note the channel reuse factors. This is a complex function of many things. For example,

- the ceiling height of the sector being used; i.e. how far an aeronautical station transmission can 'overspill' into where the frequency is reused;
- the type of service to be operated; i.e. is it protected or not; is it data broadcast such as airport terminal information service (ATIS) or ATC voice.

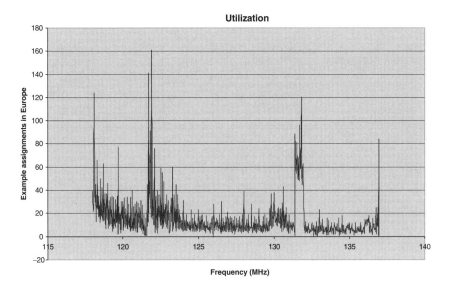

Figure 3.16 Channel utilization.

3.5 VHF 'Hardening' and Intermodulation

3.5.1 Receiver Swamping

The spectrum in the band below 108 MHz is allocated to the broadcasters exclusively (between 88 and 108 MHz). Typically the systems deployed here use very high power transmitters (typically kW) with omnidirectional antennas. Aircraft can frequently come within a few hundred metres of these, particularly when coming into land over a congested city. (Arguably the time when strong communications and navigation capability is needed the most.)

Looking at the free space path loss Equation (2.15) from the Theory section:

$$L_{\text{fspl}} = 32.44 + 20\log_{10}(d_{\text{km}}) + 20\log_{10}(f_{\text{MHz}}) \tag{2.15}$$

The term $20\log_{10}(d_{\text{km}})$ can be very small – in fact negative for ranges under a kilometre.

In addition, consider from the Theory section, the link budget equation

$$P_{\text{Rx}} = P_{\text{Tx}} - L_{\text{f1}} + G_1 - \text{fspl} + G_2 - L_{\text{f2}} \tag{2.43}$$

Again for broadcasting transmitters, the term P_{Tx} can be incredibly large (for the high powers).

Considering both these factors, it is possible for a poorly discriminating (i.e. a fairly wideband receiver filter characteristic on the effected receiver) receiver to be 'swamped' by a high-power signal in comparable amplitude to the wanted signal (Figure 3.17).

This out-of-band phenomenon was identified as a problem to avionics, and more so to the navigation equipment in the 108.000–117.075-MHz band but to a lesser extent to the communication avionics sitting above this. However the manifestation of this 'interference' is much easier to see in the communications equipment. By providing a tighter RF filter characteristic recivers can be protected from this phenomena. This is some time called 'VHF hardening'.

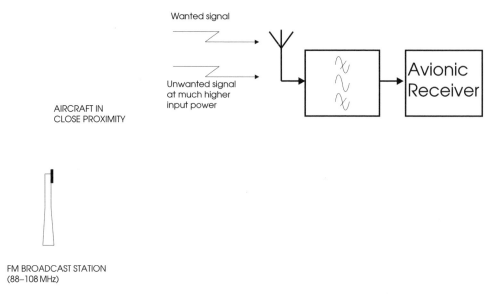

Figure 3.17 Aeronautical receiver swamping.

3.5.2 Intermodulation

Whilst the out-of-band swamping problem can in some cases be protected by the wideband filter, theoretically passing frequencies only between 118 and 137 MHz, a secondary problem was also the high-power intermodulation products produced by multiple transmitters and their transmission lines and antenna systems. This unfortunately can come straight in through the receiver pass-band filter as the products are within the 118–137-MHz range.

Recapping on the theory of intermodulation products provided in the Theory section of this book, as well as high powers and non-linear systems such as antennas, diodes, waveguide discontinuities, etc., it is very easy to generate harmonic components from broadcast and avionic multiple carrier frequencies that would fall inside the receiver filter characteristic. Theoretically, these powers would be well down on those received on the pure broadcasting tone; however, they do have the nasty characteristic that no filtering (at broad receiver input) or attenuation from channel selecting (low-pass filter in the demodulation circuitry) would be applied, and these signals can interfere and wipe out the ability to demodulate wanted signal.

As a fix to resolve both these issues (out-of-band swamping and on frequency intermodulation products) produced by the broadcasting system, ICAO introduced a desensitizing specification or hardening of the avionic receivers to this broadcasting band (Table 3.11).

3.6 The VHF Datalink

3.6.1 Limitations with VHF Voice

As the technology to support data applications became readily available in the radiocommunications industry in the late 1980s and as avionics became more computerized, it became apparent that an aeronautical mobile datalink system could be realized and moreover there

Table 3.11 ICAO FM hardening policy (see Annex 10, Vol III, Part II, 2.3.3).

- From 1 January 1998, VHF airborne receivers shall not be desensitized by FM broadcasts, when the FM signal is found to arrive at the receiver input of –5 dBm.
- Also regarding intermodulations, 'after 1 January 1998, the VHF communication receiving system shall provide satisfactory performance in the presence of two-signal, third-order intermodulation products caused by the FM broadcast signals having levels at the receiver input of –5 dBm.
- This tends to be a greater threat to on-board navigation equipment operating in band 108.000–117.975 MHz, but can be problematic for communication equipment if power is high enough.
- Prior to this policy, there were a number of modifications made to communication and navigation avionics. Those that comply with the policy are often referred to as FM hardened.

---

was a clear market for it. It could be used to pass routine air traffic services messages such as weather and pressure information, ATIS and take-off clearances. It could also be used to pass engineering information (AOC) between aircraft and their fleet, to provide the first connectivity between ATC host computers and aircraft and potentially much more.

It also provided an opportunity to introduce encryption and secure communications. It was obvious that this digitization of aeronautical mobile communications could lead to improved automation with likely consequential operational safety improvements and improved economies for the aeronautical industry.

Thus a clear market opportunity was born for data applications – an opportunity to move some of the more routine, repetitive, superfluous, verbose and cumbersome voice functions traditionally conveyed over the air traffic controller voice channel to a datalink. For some applications the datalink offered an opportunity to reduce the scope for error of interpretation, reduced air traffic controller workloads and it also could act as an affirmation channel for voice instructions (e.g. take-off clearances). It was semi-independent to the voice channel, and arguably is not as vulnerable in terms of information security.

3.6.2 The History of Datalink

In the late 1970s, some experimentation was made using basic modems operating in open mode over the standard VHF 25-kHz AM(R)S air interface. This gave rise to a number of proprietary non-standardized systems that were used by individual airlines and organizations. The first such system noted to emerge was to be the aircraft communication and addressing system (ACARS), sometimes known as VDL0 or VDLA (i.e. when the protocol is transported over VHF in the open mode described). This soon became the industry *de facto* standard and was to be the first generation of datalink systems (Figure 3.18).

It was recognized by the ICAO that standardization was required for the next generations of global data interface. VDL1 and VDL2 were planned to be standardized and were incorporated in ICAO SARPs Annex 10 in 1997. They followed shortly after the ACARS. VDL1 was standardized in case the superior VDL2 was not ready in a realistic time frame, and as it turned out, VDL1 was never implemented. VDL2 is currently under implementation (mainly in Europe). VDL3 was an American conception, also capable of carrying digitized voice as well as data and being able to operate in real time. It has been standardized in ICAO since 2001, and some experimental test systems have been built but this is still to be seen whether it will be widely adopted. Finally VDL4 has been recently standardized in ICAO (also 2001), and

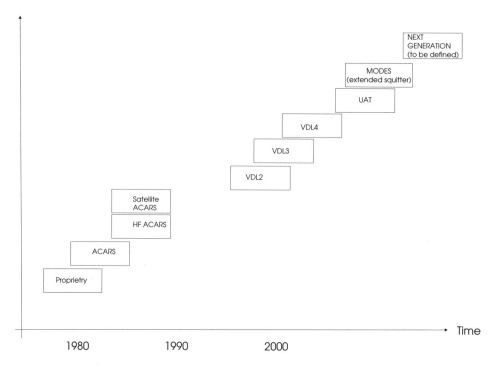

Figure 3.18   Datalink evolution.

this standard is in theory ready for deployment as pen hits paper. It is a data-only service with some quite sophisticated features enabling it to be used flexibly for surveillance, navigation and communication functions.

All the modes can act as a sub-network to the aeronautical telecommunications network (ATN), transporting data between aircraft, ATC, ATC host computers and airline computers as necessary (Figure 3.19).

3.6.3   System-Level Technical Description

3.6.3.1   ACARS/VDL0/VDLA

As the first sizable air–ground data radio solution, and still very much in practice today ACARS was available in the late 1970s. It is non-standardized in ICAO. It uses character-oriented data (ASCII based; certain characters cannot be used as they are reserved for datalink control). The air interface runs at 2.4 kbps in 25 kHz, using amplitude-modulated minimum shift keying (see Theory section; Modulation subsection). This gives a maximum payload data throughput rate of up to 300 bps (due to the primitive protocol). It can be analogously compared to a fax machine operating on a local telephone line.

Initially, services such as DCL (departure clearance), D-ATIS (digital airport terminal information service) and OCM (oceanic clearance datalink service) were migrated over from voice exchanges to routine data exchanges on this service. The detailed specification for the ACARS protocol and air interface is described in the AEEC[8] 623 specification. The standardization is

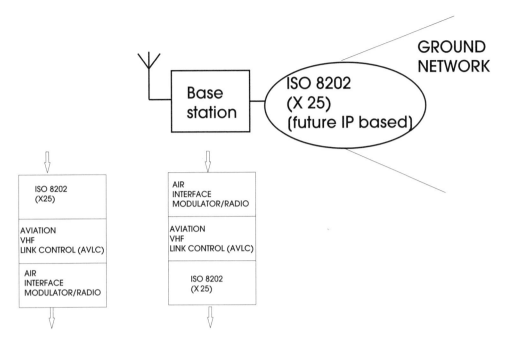

Figure 3.19   Datalink sub-network.

described in EUROCAE[9] (in Europe with three documents for the three services) and RTCA (in America).

Today there are two service providers ARINC and SITA[10] and both have 25 kHz allocations in Europe and the United States. Both services are still in service and are approaching saturation. Some migration work onto new channels has recently been carried out to re-plan the upper end of the VHF communication band and make way for its successors (Table 3.12). These have been agreed by ICAO to be the ACARS plan.

The quality of service is quite poor by communications standards, with typically over 5 % of information not being delivered.

The plan is that ACARS over VDL0/VDLA will ultimately be superseded by VDL2, VDL3, VDL4 and/or complemented by its satellite service provision and HF equivalent, which is discussed later. The VDL and satellite version have a better guarantee of delivery.

3.6.3.2   *VDL1*

VDL1 was basically conceived as a replacement for the ACARS VDL0 service, addressing its shortcomings in the quality of service. It was also planned to be the first ICAO standardized system for datalink. VDL1 was to use the same AM-ASK protocol as ACARS/VDL0/VDLA, with its low expected throughput. This was clearly the downfall of VDL1. Ironically VDL1 was standardized but was never implemented as it was overtaken by technology and the superior VDL2 before it got implemented. It is defunct.

Table 3.12 ACARS channel allotments (worldwide).

Frequency (MHz)	ARINC	SITA
129.125	United States and Canada	
130.025	United States and Canada secondary	
130.425	United States	
130.450	United States and Canada additional	
131.125	United States additional	
131.450	Japan primary channel	
131.475	Air Canada company channel	
131.525		From Sept 2004 in Europe secondary
131.550	Worldwide channel	
131.725	Primary channel in Europe	From Sept 2004 in Europe
131.825	*From Sept 2004 main ARINC in Europe*	
131.850		New European channel
136.700	United States additional	
136.750	United States additional	*From Sept 2004 main SITA in Europe*
136.800	United States additional	
136.900		Was ACARS vacated for VDL2
136.925	Was ACARS vacated for VDL2	

3.6.3.3 VDL2

The VDL2 standard was derived and agreed within ICAO in late 1996 and incorporated into Annex 10 in 1997. It was standardized in parallel with VDL1 but has a superior modulation scheme (8 DPSK) and consequently expected data throughput rate (typically a ten-fold increase for the same RF channel).

It also requires a 25 kHz slot. Today it is operational in Europe, the United States and Japan.

It is backward compatible with the ACARS service (avoids changing the ACARS data interfaces) and also supports additional data exchanges for higher speed AOC, controller pilot datalink communications (CPDLC) and some of the more recent higher data speed Link 2000+ applications (see Eurocontrol[11] website www.eurocontrol.int). Presently, in Europe it has gone into operation (since 2002) on one channel (136.975 MHz), with a plan for at least three more channels by 2008.

3.6.3.3.1 VDL Migration

Table 3.13 shows the high-level migration plan for Europe over the next few years. In summary, the ACARS slots previously described will be reduced back to five channels. VDL modes 2 and 4 will have four channels and two channels in the VHF communication spectrum, respectively (118–137 MHz). (It should be noted that VDL mode 4 will also have some allocations in the 108–118-MHz radionavigation spectrum, which is allowable because of its applications.)

Also of importance is the fact that VDL2 and 4 need adjacent 25 kHz channels to be protected. This is an oversight of the VDL2 and VDL4 design. It was discovered with testing that the adjacent channel could not be used for voice, so spectrally VDL2 and 4 could be said to require 50/75 kHz, including these 'guard bands'. The service provision in Europe again is by ARINC and SITA.

In the American region, the plan is to offer SITA and ARINC services on communal frequencies with Europe. The exact allotment is still being defined.

3.6.3.3.2 VDL Capacity

From experimental simulations carried out by Eurocontrol, Europe expects to be able to operate up to 1220 flights operating AOC only or 670 flights operating Link 2000+ (distributed over a number of base stations but with the one frequency channel).

In Europe another two channels came into service by 2005 and a total of four channels are envisaged to be in operation by 2008. This will enable a cellular frequency reuse plan and a fourfold increase in capacity. There are some migrations at the top of the VHF band required to facilitate this described in Table 3.13.

Table 3.13 VDL2 and VDL4 implementation plan for Europe.

	Frequency MHz																
	136.575	136.600	136.625	136.650	136.675	136.700	136.725	136.750	136.775	136.800	136.825	136.850	136.875	136.900	136.925	136.950	136.975
VDL 2						▨	■	■	■				■	▨		▨	
VDL 4										▨	■	▨		■	■	▨	

Source: ICAO FMG/10 meeting 2006, plan for 2008 onwards

■ channel to be used for VDL
▨ guard bands (some services may stay)

3.6.3.3.3 VDL Technical Specification

Receiver Sensitivity. As with digital systems, the receiver sensitivity is defined as the point at which the BER is 10^{-4} after forward error correction (FEC) or 10^{-3} before FEC. In the ICAO SARPs this is specified as (a) air to ground Rx sensitivity = 20 µV/m = −93.9 dBm (through an omnidirectional no-loss isotropic antenna and feeder system); (b) ground to air Rx sensitivity = 75 µV/m = −82.4 dBm (through an omnidirectional no-loss isotropic antenna and feeder system).

Air Interface. The channels are simplex 25 kHz but guard bands are required as described previously. The modulation chosen is 8-differential phase shift keying (8DPSK) running at a symbol rate of 10 500 baud. This gives a raw data throughput of 31 500 bps and better 'optimization' of the available channel compared to ACARS VDL1 and VDLAM-ASK. Obviously the actual application data rate is much less than this as overhead must be deducted. (This can be easily calculated given the frame structure) (Figure 3.20). This is converted to the X25 packet data structure at the ground station or at the aircraft terminal (Figure 3.21).

Call Set Up. Carrier sense multiple access (CSMA) is the protocol used for attaining a connection (see Theory section for description). This exploits the statistical nature of calls being required.

The latency quality specification defines delays not greater than 3.5 s for 95 % of the time for data packets. This means the protocol is not suitable for real-time, e.g. voice. It requires

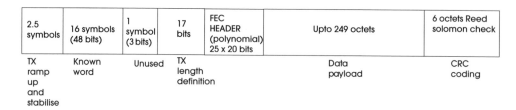

2.5 symbols	16 symbols (48 bits)	1 symbol (3 bits)	17 bits	FEC HEADER (polynomial) 25 x 20 bits	Upto 249 octets	6 octets Reed solomon check
TX ramp up and stabilise	Known word	Unused	TX length definition		Data payload	CRC coding

Figure 3.20   VDL2 air interface frame structure.

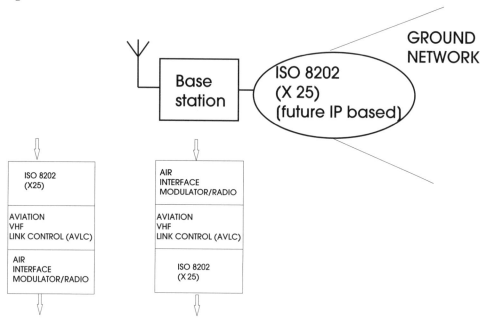

Figure 3.21   System block diagram VDL2.

the whole message being ready, i.e. store and forwarding it. This is called aperiodic traffic. The access mechanism is not organized, but randomized so any user can transmit at any time. It does not support prioritization of traffic.

In retrospect, the adoption of the CSMA protocol for VDL2 was a mistake and one of the limiting factors of the VDL mode 2 network performance. Critical ATC messages can be potentially blocked by non-critical low-priority information.

Error Control.   The datalink is controlled by the CSMA and embedded in the data packet are the block coding and error correcting polynomials. The datalink control analyses this and performs the necessary error correction.

Resend Protocol.   Also embedded in the protocol is for the receiver to receive a packet of data and acknowledge it back to the transmitter. Messages can be resent a number of times (with varying repeat times) to improve the quality of service (when compared to ACARS, which does not have this facility).

The frames are interleaved extensively to protect from multipath fading events plus the input data is scrambled (for easy clock recovery).

Handover Mechanism. At the boundary between cells (i.e. as the aircraft is being handed over from one base station to the next) there are four mechanisms that can be set in the VDL protocol for triggering this:

• Monitoring signal quality parameter in the data frame;
• timer expired of the retry events;
• N2 degraded link up;
• N2 degraded link down.

For more details, see the ICAO VDL mode 2 manual.

3.6.3.3.4 VDL2 Co-Site Issues

One of the main issues with VDL2 is its compatibility with on-site (i.e. on the same aircraft) DSB-AM voice (and also applicable to ACARS or other services in the VHF band) (Figure 3.22).

Here at least two services are operating in the same VHF band and therefore the filter characteristic of the VDL receiver in reality does not filter out the transmitted voice signal sharply enough and vice versa. When analysing the link budget, because the free space path length is just over a few metres (maximum 30 m) and the launched transmit power is usually 10–20 W, the unwanted voice signal swamps the VDL receiver; the problem exists in reverse as well. This is one of the reasons that VDL and voice allotments are in separate portions of the VHF communication band and there is a future migration plan for this.

From a practical point of view, it is impractical to interleave voice and data transmissions, and there are times when they are both concurrent. At such times as this, the usual data glitch can be heard by a pilot or the voice wipes out the ability to properly decipher the data and it needs to be resent. Some modelling of this work has been carried out by Eurocontrol[12] and the

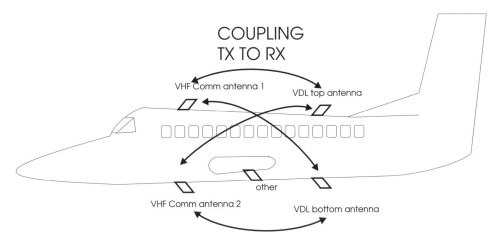

Figure 3.22 Co-site issues for VDL.

Table 3.14   ICAO co-site policy from Annex 10.

If there are DSB-AM voice and VDL on the same aircraft, care needs to be taken when sighting the radio equipment and, in particular, the antennas to stop either system interfering with the other. ICAO has defined in SARPS that 'the receiver function shall provide adequate and intelligible audio output with desired field strength >150 μV/m ($= -102$ dBW/m^2) and with an undesired VDL signal field strength of at least 50 dB above the desired full strength of any assignable channel, 100 kHz or more away from assigned channel or designated signal'.

conclusion is that this annoyance can be tolerated by both data and voice users, provided they are separated by at least one 25 kHz channel. In addition, in a busy sky with high PIAC and consequently a heavy-loaded VDL2, the retransmissions due to data/voice collisions can be kept round about the 2 % mark. The same problem exists between other services but usually they are in different bands and the front-end filter on the receiver can provide enough rejection (Table 3.14).

3.6.3.3.5   VDL Receiver Filter Characteristic

This is a pulse shaping raised cosine filter of alpha $= 0.6$ to reduce as much as possible spurious in adjacent channel. The emission designator for licensing purposes is 14KG1DE.

3.6.3.3.6   Interaction to the Aeronautical Telecommunications Network

VDL2 uses standardized protocols to enable the air interface (the VDL2) to plug directly into conventional packet database structures. The ISO 8208 standard packet switch network is supported by X25 and in the future will migrate to an IPv6 network. Figure 3.21 shows how data connectivity can be achieved between end user (the pilot, controller or airline computer).

3.6.3.3.7   Problems with VDL2

As discussed, the effects of adjacent bands on the VDL 25 kHz channel and also the inverse as well as the susceptibility of the VDL2 signal to signals in the adjacent band had led initially to a requirement to sterilize the adjacent band for good functioning of the VDL2 system. Obviously, with the VHF spectrum becoming thoroughly congested, this is a less than desirable status and ultimately points to a deficiency in the VDL2 design. In 2000, it was proposed to tighten the VDL2 specification in particular with regard to adjacent channel emissions. This has been accepted and has come into effect.

3.6.3.4   *VDL Mode 3*

3.6.3.4.1   VDL3 History and Politics

In the 1990s the United States led the charge with moving to a digitized voice service. This led to an ICAO COM/OPS/DIV[13] decision to implement digital time division multiple access (TDMA) base voice and data as a datalink in the VHF band. Hence VDL3 was conceived.

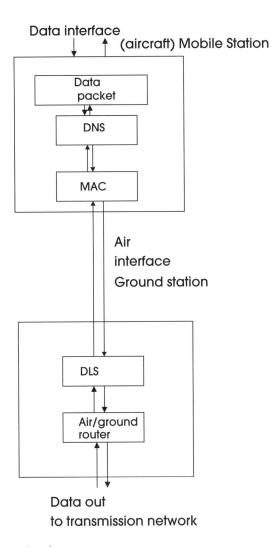

Figure 3.23   VDL2 data interface.

Further in the late 1990s, the United States decided not to go with 8.33 channelization proposed by Europe and for the first time in ICAO, two separate long-term strategies were in place to solve the VHF traffic growth and congestion problem.

The VDL3 TDMA frame structure, which operates in real time and hence has the ability to carry voice, has a system architecture and network topology very similar to GSM mobiles, which use the same TDMA access technology.

To date there is no widely deployed network and decision to go with it is back on hold. In the United States, some experimental or pilot networks have been constructed with some experimental and validation testing being carried out by Mitre[14]. In Europe, currently there are no plans to deploy VDL3.

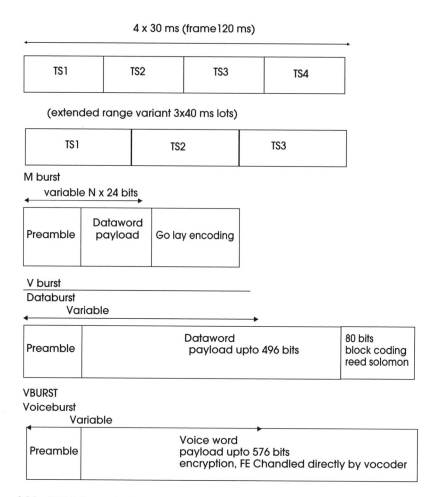

Figure 3.24 VDL3 frame structures.

3.6.3.4.2 VDL3 Technical Specification

VDL3 is the first aeronautical communications system to use TDMA (see Theory section). The actual format of the air interface is a four-slot TDMA system. Each slot can be configured as a voice channel or a data circuit. Frame rate is 1/120 ms and slot rate is 1/30 ms.

The D8PSK modulation is running at a raw speed of 10.5 baud (symbols per second), which is equivalent to 31.5 kbps (Figures 3.24 and 3.25).

VDL3 Link Control.   The downlink (D/L) (air to ground) is used for link access and status. The uplink (U/L) is used for timing synchronizing and configuration. A timing advance is required as a function of distance from base station. This is to ensure that the message arrives in allocated time slot and is slotted in with other mobile transmissions with no overlap.

VDL3 Sub-Data Rates: Management of Sub-Channel/Voice/Data Sub-Channel.   Within the time slot, a management burst is used as a preamble to the payload. This is the overhead

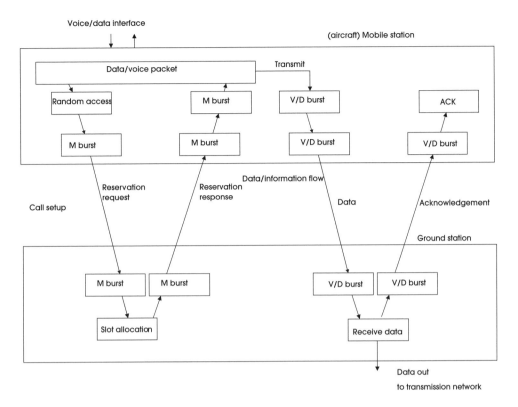

Figure 3.25 VDL 3 protocol.

controlling the air interface (synchronization, timing, call set up, type, etc.), and is called the 'M' burst. Immediately following this is the payload burst carrying the data or encoded voice, called the 'V' burst.

VDL3 M Burst. The M burst is a variable-length burst of $n \times 24$ bits. Golay encoding is used with 12 bits of coding to 12 bits of data to improve BER under path fading, distortion or self or external interference and to make the management information robust and reliable.

VDL3 V Data Burst. For data, there can be up to 496 bits of data in the payload. To this a Reed–Solomon encoding polynomial is applied. This adds a further 80 bits of block coding overhead. This can correct up to 40-bit errors in the data stream. No data interleaving is used on VDL3 to counteract fading conditions, and it works on a re-send technique instead. So this gives a raw data speed of V (data) = up to 496.000/0.120 = 4.13 kbps. (This is matched quite closely to a typical CPDLC frame size equal to typically 500 bits, including 150 % overhead).

VDL3 V Voice Burst. With voice, it is encoded in a proprietary codec using an 'advanced multiband excitation vocoder' (AMEV) technology. This operates with a payload of 576 bits per frame raw voice. So this gives a user interface speed V (voice) = up to 576.000/0.120 = 4.8 kbps invoice sample rate (note camparison to 48 kHz analogue chanels).

The voice bursts are handled directly by vocoder to optimize the voice encoding (or compress good-quality speech into this minimal bit rate.) FEC is used by the vocoder. Encryption can also be added at the vocoder.

VDL3 Extended Range. Ultimately VDL3 becomes a range-limited protocol, and to overcome this in some of the rural cells, a three-slot TDMA structure with extended guard bands can be used. This is a common technique in TDMA. It is also used in GSM where the conventional maximum GSM cell range of 37 km can be doubled by changing from an eight-slot TDMA to a four-slot TDMA or 'Dumbo cell' structure.

The extended range system uses 40-ms time slots with extra 10 ms apportioned to guard, extending bands for each slot. In theory this extra 5 ms each side of burst extends the range of the cell to an additional 1500 km, making the horizon the limiting factor again.

VDL3 Service Combinations. The VDL3 frame structure allows flexible combinations of payload. They can be, for example, all voice or all data. Table 3.15 describes the possible combinations.

Voice services support prioritization, pre-emption and digital services such as call waiting, priority, etc.

The data interface at the ground and to the avionics is based on ISO 7 layer model using interdomain routing exchange protocol interfaces with ATNs. The address format is XXXYYZZ. (ZZ is aircraft code, XXX and YY are airline locators.)

Table 3.15 VDL3 service combinations.

Standard range	4V
	3V1D (usual configuration)
	2V2D
	3 Flexible circuits
Extended range	3V
	2V1D

3.6.3.5 VDL4

3.6.3.5.1 VDL4 History

VDL4 was originally conceived and planned as a broadcast link used to support navigation and surveillance functions. It was initially proposed to ICAO by European interests in 1994, and standardized by SARPS in 2000, EUROCAE MOPS for ADS-B in 2001 and by ETSI[15] in Europe in 2002.

VDL4 was designed to provide for the deficiencies in VDL0, 2 and 3. It supports message priority handling and has the potential to support time-critical communication services and further AOC applications. It can also work in point-to-point mode as well as the point-to-multipoint (broadcast) mode for which it was designed.

It was also geared more towards surveillance functions.

3.6.3.5.2 VDL4 Technical Specification

VDL4 Air Interface. The air interface deploys Gaussian frequency shift keying (GFSK) with a data rate of 19.2 kbps. GFSK optimizes information rate versus the *C/I* (carrier power to interference power), which provides the best possible spectrum efficiency in bits/Hz. This means that the frequency reuse coordination is lowered. There is no FEC deployed in the air interface. This is done by the application level.

The protocol is a new 'self organizing time division multiplexing' (STDMA). It is a more considered and intelligent form of organizing random access to the timeslots between the many users. It is a statistical way of reserving and prioritizing allocation between the typical users. It supports prioritization of payload. It provides the broadcast of automatic dependent surveillance (ADS-B), thus extending the concept just beyond ATC data to cater for navigation and surveillance functions.

It has a long TDMA structure with a superframe (60 seconds) with 4500 slots (each 13.33 ms) (Figure 3.26). Reservation of timeslots within this frame by applications can be made up to 4 minutes in advance. Each slot can be used for a transmit or receive signal by any mobiles operating in the 'net'.

Aircraft course position is broadcast at least once per minute (can be broadcast more often, e.g. every 5 seconds as required).

The self-organizing concept (each application reserves a slot) allows VDL4 to operate efficiently without a need for a coordinating station. This is implemented using a central control channel called 'global signalling channel' (GSC). The GSC allows each mobile to work autonomously. Thus no ground infrastructure is actually required by VDL4, although a ground station can be used to control the network if preferred or a ground station can be configured to act just like another VDL4 mobile. This makes VDL4 particularly flexible and useful over oceanic or polar travel for aircraft to aircraft communications, navigation and surveillance functionality.

There is a prioritization stack for allocating slots in the frame. An oversimplification of this would be to consider four stages of priority: slot empty, slot used, slot used for high-priority point-to-point and slot used by highest priority broadcast. A more detailed description can be found in the ICAO VDL4 manual.

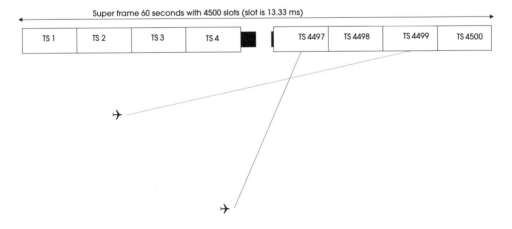

Figure 3.26 VDL4 superframe.

VDL4 Synchronization. Synchronization of each of the VDL4 mobiles to universal standard time can be performed by a number of means – again a priority stack exists for the synchronization and are in order of priority: via satellite receivers such as GPS or in future Galileo, atomic on-board clocks, synch from ground networks (when available), synch from other mobile users or ultimately floating (or non-synched operation).

VDL4 Hardware. Minimum ICAO MOPS specification requires one transmitter and two receivers per mobile (aircraft).

3.6.3.5.3 VDL4 Channelization

In Europe, VDL4 presently has one channel allocated in VHF communications band (currently 136.950 MHz, but in the process of changing to 136.925 MHz to fit in with the deployments of VDL2); see again Table 3.13. There is a second channel planned in Europe for 2008 (136.825 MHz).

A provision in the navigation band for VDL4 (under WRC 2003) to support 'ATN compliant sub-network services for surveillance purposes' was made. Theoretically, this means that VDL4 can be allotted anywhere in this band. The practical realities are that this band is similarly congested in core flight areas with VHF omni range equipment and instrument landing system assignments and is planned to accommodate the future ground-based augmentation system to complement satellite navigation, so currently there is only a plan for one provisional VDL4 channel in this band.

There is presently no VDL4 planned for the North American region other than autonomous use over the oceanic/polar region on an airline-by-airline basis.

3.6.3.5.4 VDL4 Services

VDL4 supports many data services ATC and non-ATC (e.g. advanced surface movement guidance and control system, surveillance function).

3.6.4 *Overview of the Modes – A Comparison*

Tables 3.16 and 3.17 summarize each of the VDL modes.

3.6.5 *Services over Datalink*

In a world of ever digitalizing technology, there are a growing number of data applications: some support the primary function of ATC, others support non-ATC core functions but are still required for the safe and efficient operation of flight.

Some of the applications available today are listed in Tables 3.18 and 3.19, together with their acronyms and it is also indicated which datalink modes can handle them.

3.6.6 *Future Data Applications*

This is an area seeing incredibly fast growth over the last few years. It strengthens the hypothesis that once a system becomes available, i.e. VDLs 2, 3 and 4, the users will take to it and demand can fast outstrip supply.

Table 3.16 Overview of VDL modes.

	ACARS/VDL0/VDLA	VDL1	VDL2	VDL3	VDL4
Standardization complete	Not standardized	NA	1997	2000	2000
In operation	Since late 1970s	Obsolete	2002	Experimental networks since 2003	Imminent
Modulation	AM-ASK	AM-ASK	8DPSK	8DPSK	GFSK
Air interface format + protocol	Unstructured	CSMA	TDMA-CSMA	TDMA	S-TDMA
Air interface speed	2.4 kbps	2.4 kbps	31.5 kbps	31.5 kbps	19.2 kbps
Payload (speed)	300 bps	300 bps	>10 kbps	Up to 4 × 4.8 kbps (19.2 kbps)	<19.2 kbps
Prioritization	Not supported	Not supported	Not supported	Supported	Supported 4 levels
Data services supported	✓	✓	✓	✓	✓
ATC (DCL, D-ATIS, OCM)	✓	✓	✓	✓	✓
AOC			✓	✓	✓
Link 2000+					
Navigation and Surveillance functions					✓
Voice services supported	No	No	No	Yes	No
Service providers	ARINC/SITA	ARINC/SITA		SITA	
Channelization	25 kHz + guard bands	25 kHz + guard bands	25 kHz + guard bands	25 kHz	25 kHz + guard bands
Deployment region	All	None	All	North America	Europe
Channel capacity (users)	Typical 600+ per channel	Typical 600+ per channel	Dependent on coverage volume	Theoretically up to 4500 users, not validated	
Rx sensitivities					

Table 3.17 Strengths and weaknesses of VDL modes.

	ACARS/VDL0/VDLA	VDL1	VDL2	VDL3	VDL4
Strengths	Resilient Proven		Standardized Moderate speed	Voice and data on one radio Supports prioritization and time-critical applications	Efficient payload data throughput Flexible applications (+ nav + surv)
	Mature		Applications tested	More efficient payload throughput than ACARS/VDL2	Air–air communications
	Deployed		Proven		Self-managing, good for remote areas
Weaknesses	Limited speed and application	Obsolete	Deployed Co-channel problems	Not fully tested, validated	Validation testing not complete
	Saturated Non-standardized		No prioritization of traffic No voice or real-time handling	Late implementation as yet Commitment to deploy	Does not support real-time voice Single point of failure for multiple systems
			Weak protocol	Single point of failure for data and voice	Not implemented as yet

Table 3.18 VDL services.

		Acronym	ACARS/ VDL0/VDLA	VDL1	VDL2	VDL3	VDL4
ATC	Automatic dependent surveillance	ADS					✓
		CM					
	Controller pilot datalink communications	CPDLC			✓	✓	✓
	Digital flight information services/automatic terminal information service	D-FIS/ATIS	✓	✓	✓	✓	✓
Non-ATC function		AIDC					
		AMHS					
	Cockpit display of traffic information	CDTI			✓	✓	✓
	ADS broadcast B mode	ADS-B					✓
	ADC broadcast C mode	ADS-C					✓
	Flight status	FS	✓	✓	✓	✓	✓
	Aircraft situational awareness	AIRSAW					✓
	Engine performance		✓	✓	✓	✓	✓
	Fuel status		✓	✓	✓	✓	✓
	Crew identification		✓	✓	✓	✓	✓
	Weight and balance		✓	✓	✓	✓	✓
	Off, out, on and in times	OOOI	✓	✓	✓	✓	✓

CM, Context management; AIDC, ATS interchange data communications (where ATS stands for air traffic services); AMHS, ATS message handling services.

Table 3.19 Typical VDL service data lengths (kilobytes).

Service	Airport U/L	Airport D/L	TMA U/L	TMA D/L	En route U/L	En route D/L
FIS	0.2	0.0	0.9	0.0	6.9	0.0
TIS	23.7	0.0	7.0	0.0	20.5	0.0
CPDLC	3.4	2.9	1.3	0.9	1.1	1.3
DSSDL	0.2	0.3	0.1	0.2	0.1	0.1
AOC	0.4	8.4	0.6	8.5	0.2	3.5
ADS reporting	0.0	16.1	0	3.3	0.0	1.5
AUTOMET	0.0	0.0	0.0	4.4	0.0	6.2
APAXS	0.0	0.0	0.0	0.0	131.7	115.5

Further Reading

1. The 'Aeronautical Radio Incorporated' website, www.arinc.com
2. See International Civil Aviation Organization (ICAO) website, www.icao.int
3. Aeronautical Mobile (Route) Service as described in the International Telecommunications Union Radio Regulations, see www.itu.org

4. Frequency Management Group of ICAO Europe, based in Paris. www.paris.icao.int
5. ECAC. European Civil Aviation Conference. www.ecac-ceag.org
6. International Civil Aviation Organization (ICAO) Standards and Recommended Practices (SARPs) Annex 10, Chapter 2
7. Okumura–Hata Propagation prediction model, *Trans.Vehicular Technology*, VT-29, pp. 317–325, IEEE 4980
8. Airlines Electronic Engineering Committee (AEEC). See www.aviation-ia.com/aeec
9. European Organization for Civil Aviation Equipment EUROCAE
10. Societe Internationale de Telecommunications Aeronautiques (SITA)
11. European Organization for the Safety of Air Navigation (Eurocontrol) see www.eurocontrol.org
12. Eurcocontrol Co-Site Testing, see www.eurocontrol.be/VDL2/public/standard page/cosite.html
13. ICAO COMS/OPS/DIV. www.icao.org
14. Mitre Corporation, see www.mitre.com
15. European Telecommunications Standards Institute (ETSI), see www.etsi.org

4 Military Communication Systems

Summary

This chapter looks at the military communication system legacy as it was first deployed in the VHF band immediately above the civilian system. It goes on to explore the reasons for a more secure, robust and survivable communications systems and how this leads to the joint tactical information distribution system (JTIDS) and later the multifunctional information distribution system (MIDS) in the UHF band.

4.1 Military VHF Communications – The Legacy

Prior to International Civil Aeronautics Organization (ICAO) in 1947, mobile aeronautical communications for civil or military use had a similar history, similar architecture and the same equipment. It was in the 1940s that military communications became proprietary, secret and clandestine for all the obvious reasons. With the advent of ICAO and the new ITU radio regulations that separated aeronautical mobile (route) service from the aeronautical mobile (off-route) service, (AM(R)S and AM(OR)S respectively) the two paths were to diverge forever. Immediately after the divergence of paths, the military systems employed were initially much the same as their ICAO counterparts and as such just a natural extension to this, above the 137-MHz line.

The discussion in Chapter 3 of this book, looking at generic VHF aeronautical communications, can largely be extrapolated up to the 'off route' band, which runs from 138 to 144 MHz in most countries. This is still largely used for more routine (non-combat or mission critical) military movements today, with this higher band being largely a military overlay that sits in duplicate above the civil function. At many air traffic control centres, there are military control rooms mirroring the functionality of the civilian systems with a few minor differences.

Aeronautical Radio Communication Systems and Networks   D. Stacey
© 2008 John Wiley & Sons, Ltd

The channelization of this band is 25 kHz; there are no plans to go to 8.33 kHz working. The channels are not exclusively used by the military; there is sharing in this band with fixed mobiles, maritime mobiles and land mobiles and also space to earth, sometimes on an equal 'primary' basis. The exact arrangement varies from state to state.

4.2   After the Legacy

It should be pointed out that different militaries deploy different systems (Figure 4.1). Some of them (and certainly the latest versions and equipment) are undeclared to the public domain and some undeclared to the regulatory domain (i.e. they do not appear in the International Telecommunications Union, ITU) for the obvious reason of national security.

The following are the other military bands of note:

30–88 MHz: The SINCGARS (single channel ground and airborne radio system). This is a mature tactical system developed in the 1970s with production in the late 1980s. This is operated for secure military data and voice. The channelization of this is 2320 channels of 25 kHz. The system operates using frequency hopping. (This allocation is undeclared in the ITU but is known to be operated in at least one country.)

225–400 MHz: The 'HAVEQUICK' communication system. This mature tactical communication system was also developed in the 1970s and deployed in the late 1970s in state-of-the-art US fighter aircraft. It has full 7000 × 25 kHz channels and is also used in hopping mode. Again it has no international regulatory status.

Enhanced position location reporting system/situation awareness datalink (EPLRS/SADL): This also is a pre-planned tactical communication system, predominantly used by ground forces, but of late also installed in state-of-the-art fighter aircraft. This system was also developed in 1970s and deployed in the 1980s. Its band of operation is 420–450 MHz. It is a secure data system and is also undeclared in ITU.

JTIDS/MIDS: This operates between 960 and 1215 MHz. This will be discussed in detail as it is the current state-of-the-art military system, or the most recent one disclosed again this is not allocated by ITU.

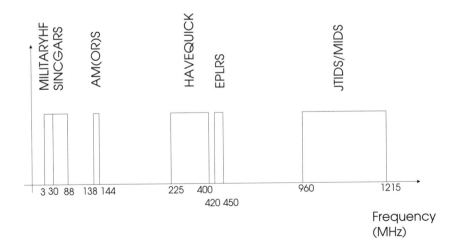

Figure 4.1   Military bands and systems.

4.3   The Shortfalls of the Military VHF Communication System

By their very nature military systems are designed to ultimately safeguard national interest from 'the enemy'. When being operated in this manner, it is imperative that the communications are secure, unpredictable, encrypted as necessary and robust to interference, large electromagnetic forces and wilful jamming. They must also be of a very high reliability and have a 'survivability' when parts of the infrastructure are possibly destroyed, etc. All this has led the world's military to develop something a bit more robust than using the VHF communication architecture still used by civil aviation and military when performing civil duties.

Traditionally, the military has had a bigger budget to develop these 'toys' for sophisticated 'war games'. It has been estimated by some that the piece-by-piece equivalent in a military system can be 10 times the cost of a conventional 'open' first mobile system, such as the VHF AM(R)S system. Certainly the technical specification is usually up by a grade or two in overall performance; it is also where the term 'military spec' has been gleaned.

4.4   The Requirement for a New Tactical Military System

The design specifications for a mobile military communication system that embraces all aspects of military requirements from surveillance, reconnaissance to combat itself and all the support requirements and scenarios in between have the following design challenges:

Security – that is encryption of voice and data.
Increased data speeds – the legacy VHF system channelization severely restricts the useful
    amount of data that can be put through one channel as can be seen in Chapter 3. When
    maximum data rates are around 9.6 and 19.2 Kb/s the military needed the highest data
    speeds practically possible over-the-air interface.
Jam resistance – that is the ability of the system to carry on under purposeful interference or
    jamming.
Reduced size of avionics and terminals – it was important to use the latest technology to drive
    this down – weight in aircraft = cost, more in terms of real estate cost and the loss of agility
    implications but also fuel burn cost.
Surveillance and navigation functions – gain properties of the new system could be used to
    triangulate mobiles with known propagation times, etc., and also participant identification.

4.5   The Birth of JTIDS/MIDS

The development of this new system was mainly led by the United States and in later stages was joined with the NATO states of western Europe. Today, JTIDS/MIDS is operated by at least 25 states, including 17 European States, United States, Canada, Israel, Iceland, Australia, New Zealand, Japan and South Korea.

The system is designed to operate in the UHF band between 960 and 1215 MHz in parallel with the aeronautical radionavigation service. However, it is important to note that (a) JTIDS/MIDS does not have a formal status under the international radio regulations, (b) there is no spectrum allocated to it as such in the ITU, and (c) it operates under multilateral agreements between the countries mentioned above. Some purists say 'It doesn't exist', enhancing its clandestine reputation. In theory, it is operated on a 'no interference basis' with the radionavigation functions (namely DME and GNSS). Usually this is achievable by the nature of where the exercises are conducted in segregated offshore military test zones, but obviously in a time of warfare the civilian airspace demand disappears around the direct area where this tactical system would be used. However, compatibility between the three systems can be maintained for specific circumstances usually detailed in national and multilateral agreements.

An evolution of system architecture has occurred since its early conception. Today, important earlier generations are Link 4A and Link 11 architecture, but the current state of the art resides with Link 16. Link 4A and Link 11 will be described briefly in passing, but deeper concentration will be given to Link 16.

4.6 Technical Definition of JTIDS and MIDS

4.6.1 Channelization

The JTIDS/MIDS spectrum between 960 and 1215 MHz is cut into 51 channels of 3-MHz separation (Figure 4.2). Frequencies within ±20 MHz of the carrier centre at 1030 MHz and 1090 MHz are not used. This is reserved for the military Identification Friend or Foe system (and also the civilian secondary surveillance radar operates here).

4.6.2 Link 4A Air Interface

Link 4A has control messages (V-series messages) and aircraft responses (R-series messages) (Figure 4.3). A 56-bit V-series message is sent every 32 ms, giving a one-way data rate of 1750 bps. The two-way effective data rate is slightly higher at 3000 bps.

4.6.3 Link-11 Air Interface

Link 11 uses 24-bit data payload per frames (plus an extra 6 bits for error correction and detection), with up to 75 frames/s fast rate. This gives an air interface rate of 2250 bps or a useful payload data rate of 1800 bps (Figure 4.4).

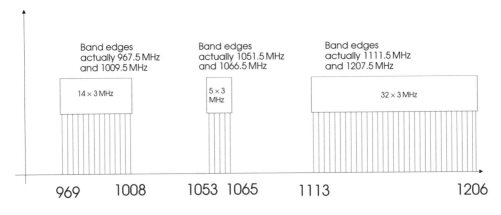

Figure 4.2   JTIDS/MIDS channelization plan and modes.

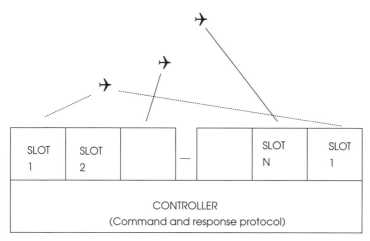

TDMA FRAME STRUCTURE

Figure 4.3   Link 4A air interface.

4.6.4   Link 16 – Air Interface

Link 16 uses aspects of time division multiple access (TDMA), Frequency division multiplexing (FDM) and frequency division multiple access (FDMA), and code division multiple access (CDMA) architectures (Figure 4.5).

The TDMA protocol is used with time slots of duration 1/128 s or 7.8125 ms. Each mobile on a system is pre-assigned to which time slots it has to use for transmitting and receiving. The frame for Link 16 is 12-second long; thus it has 12 × 128 slots or 1536 slots.

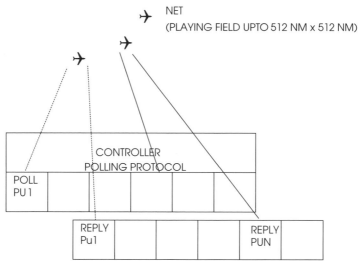

PU = P articipating unit
n.b Controller centric

Figure 4.4   Link 11 air interface.

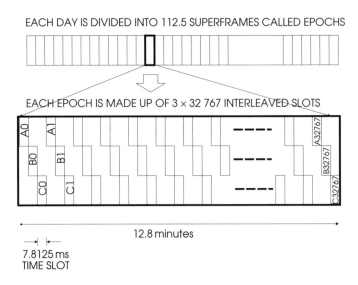

Figure 4.5   Link 16 frame.

The superframe called an *epoch* has a duration of 12.8 minute (and 98 304 slots). It has three slots of frames interleaved within it repeatedly (3 × 32 767 slots). The sequence of these slots is interleaved as follows: A-0, B-0, C-0, A-1, B-1, C-1,..., A-32767, B-32767, C-32767, and then it starts over again. (The numbers above are called the slot index.)

The FDM and FDMA protocols are used in that there is a pre-arranged (pseudo-random) algorithm for changing frequency across the 51 channels. This is done independently of the slot duration and much more rapidly at every 13-μs intervals. This is known as frequency hopping or as a spread spectrum technique (Figure 4.6). These hop patterns are provided to the mobiles ahead of them, joining a network in what is called the 'mission data load' or MDL.

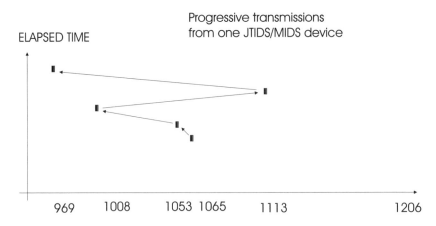

Figure 4.6   JTIDS/MIDS frequency hopping.

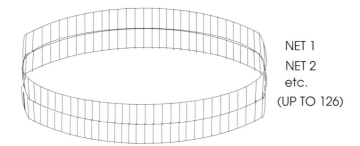

NET 1
NET 2
etc.
(UP TO 126)

Figure 4.7 JTIDS/MIDS 'net' concept.

The system can also operate in two other modes, but these are not seen so often as they are not as secure and robust:

A. With encryption and a single frequency (usually 969 MHz);
B. No encryption, single frequency (969 MHz).

The CDMA principle is applied in that the data 'payload' bitstream is transferred from the data communications domain to the radio frequency waveform domain using a pulse modulator with pseudo-random encoding. This makes the waveform more tolerant to jamming and able to operate over the same time slot and frequency as other systems with only a degradation but still the ability to be demodulated.

Another related dimension of the JTIDS/MIDS is the concept of a 'net' (Figure 4.7). This is where multiple users in TDMA mode, all frequency hopping around (in FDM mode) up to 51 channels with the pseudo-random pulse modulation (CDMA mode), can be overlaid on another different set of users, using different hopping sequences and intervals. In addition, redundant slots from one net can be used by multiple users simultaneously in another net.

Multiple nets can be 'stacked' or used in parallel in a given service volume. Here each net is assigned a number (between 1 and 128); this number defines the 'network participation group' and the unique hopping sequence. Within a network participation group, a number of attributes will be allowed to the participating mobiles. Typically up to 20 nets can be operated concurrently in a given service volume before the intersymbol interference between the nets starts to degrade the quality of the services carried. With careful planning, up to 127 nets can be operated over a wider service volume. This allows individual elements or 'battle groups' to be formed according to function and purpose, sectors of control or geographical areas.

4.6.5 Access Methods

After initial entry into a network, the sharing of the time slot resource can be apportioned in a number of different ways:

- *Demand assigned.* This is latency critical and is used to support a push-to-talk function. There can be contention if two users wish to use this at the same time.
- *Dedicated access.* This is when each of the reporting mobiles is given dedicated time slot reservations.

Figure 4.8 Link 16 message format.

- *Random access.* This is where a number of applications can 'contest' for the resource on a statistical sharing basis. Usually the receiver will vote for the stronger signal and the other signal will be discarded. The QOS therefore is a function of number of mobiles and average data traffic and the amount of resource the system sets aside for this function.
- *Time slot reallocation.* This is when mobile applications are sharing time slot resource in a pool according to their pre-booked demand expectations. The expected requirement for the next number of seconds (or even tens of seconds) is embedded in a transmission and the corresponding receiver calculates the 'time slot reallocation' on a dynamic basis and relays this back to the user pool on its next transmission.

4.6.6 Link 16 Data Exchange

Within the link layer for Link 16, there are multiple packets that can be sent in any one time slot. The general message format is a standard serial digital data stream of information. This is encoded into a number of pulses. The reason for pulsed modulation is its resilience to jamming of an intruder system (Figure 4.8).

4.6.7 Jitter

A randomized space is left before the start of the message, which is called the jitter, and the randomized element is important to minimize the probability of jamming.

4.6.8 Synchronization

This is followed by a number of synchronizing pulses and timing refinement pulses. It is worth mentioning at this point that there are a number of sources from which any one mobile can take its timing. The architecture is mainly designed so that the maximum independence can be given to any part of the network from a central control and synchronization function. This obviously ensures resilience and survivability of the system in a conflict or failure situation.

In the short-term each of the mobiles can work autonomously from its highly accurate atomic clock. From time to time it is necessary to align this with the master timing source for a network (which can be moved around from one mobile to another and is usually called the 'network time reference' or the NTR) or by taking a bearing against a secondary source which is known to be more accurate than the individual mobile. This is done using the synchronization pulses and timing refinement pulses.

4.6.9 Synchronization Stack

Primary. The NTR broadcasts transmissions of the system time and the network entry messages.
In the first instance a mobile looks out for these transmissions and uses them to synchronize.

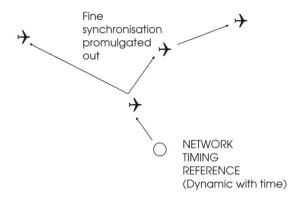

Figure 4.9 Round-trip-timing protocol.

Secondary. Once the NTR is established, units close into it and certainly within LOS can assume accurate clocks. Once in fine synchronization with the NTR, these units assume a status of 'initial entry JTIDS units' or IEJUs. These can be used as a secondary source of synchronization if the NTR is over the radio horizon from the mobile in question.

Round-trip-timing (RTT) messages can be used to attain fine synchronization. Such messages do not contain any data and as such units that are not 'finely synchronized' are discouraged from sending data messages until they establish this fine synch.

The RTT message just contains synch (including time refinement) and header portions of a message. RTT interrogations (RTT-Is) are sent at the beginning of a time slot by units attempting fine synchronization; RTT-Rs are usually sent by NTR or IEJUs in response at the end of the same time slot. By analysing the total round trip timing, the affected unit can determine the necessary correction to its timing (Figure 4.9).

Tertiary. Failing the synchronization to the primary or secondary source, the mobile unit will revert to its highly accurate atomic clock, which can be related to a GNSS timing source. The unit will be able to receive messages from other mobiles but will be unable to transmit data until it can elevate its synchronization status to primary or secondary.

4.6.9.1   Header

The message header is sent to identify the kind of data packing that is about to follow, or in the case of the RTT, that no data message is following.

4.6.9.2   Data Packing

Data is transmitted in Link 16 in the form of data blocks of 70 bits, which are called 'words'. There is a number of data formats that are used depending on the application, jamming resilience and the environment; i.e. there is a trade-off between data throughput and its robustness or resilience to jamming. For example, the packed 4 single pulse format provides the highest data throughput whilst there is minimal data duplication and protection in waveform, and therefore it is most prone to jamming. Similarly, the standard double pulse (STDP) represents the highest immunity to a jamming environment, but consequently it has the minimum net data throughput (Figure 4.10). A more detailed analysis of the pulse combinations follows.

Figure 4.10   Standard double pulse.

4.6.9.3   Standard Double Pulse Format

The data is entirely duplicated and transmitted in both data slots. This gives a high resilience to jamming. The word has a data payload of 70 bits. Up to three such words can be transmitted in any STDP format. This gives a total data throughput capacity of $3 \times 70 \times 128 = 26.88$ kbps. This is the default format for sending information as it provides maximum anti-jam ability. Only when a higher data throughput rate is required does the format step up.

4.6.9.4   Packed 2 Single Pulse Format

This has a similar format to the STDP, except the data being sent is not duplicated. Thus this format has twice the throughput of the STDP (i.e. 6 words per frame) at the cost of jamming resilience. The effective data throughput rate can therefore be up to 53.76 kbps (Figure 4.11).

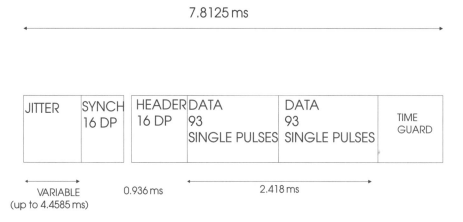

Figure 4.11   Packed 2 single pulse.

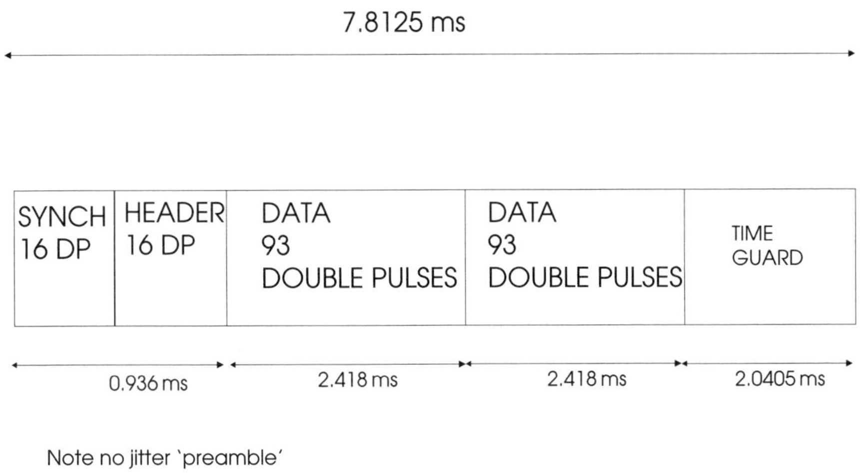

7.8125 ms

SYNCH 16 DP	HEADER 16 DP	DATA 93 DOUBLE PULSES	DATA 93 DOUBLE PULSES	TIME GUARD

0.936 ms   2.418 ms   2.418 ms   2.0405 ms

Note no jitter 'preamble'

Figure 4.12 Packed 2 double pulse.

4.6.9.5 Packed 2 Double Pulse Format

This is an alternative form of packing: It restores the anti-jamming properties by double-sending data and also increases the data throughput rate from the STDP format to 6 words per frame (or up to 53.76 kbps). The downside is a compromise on jitter period, which slightly decreases the anti-jam tolerance (Figure 4.12).

4.6.9.6 Packed 4 Single Pulse Format

This format provides the highest data rate (12 words per frame) or 107.52 kbps but with the lowest immunity to multi-path propagation and jamming due to the data being totally unduplicated (Figure 4.13).

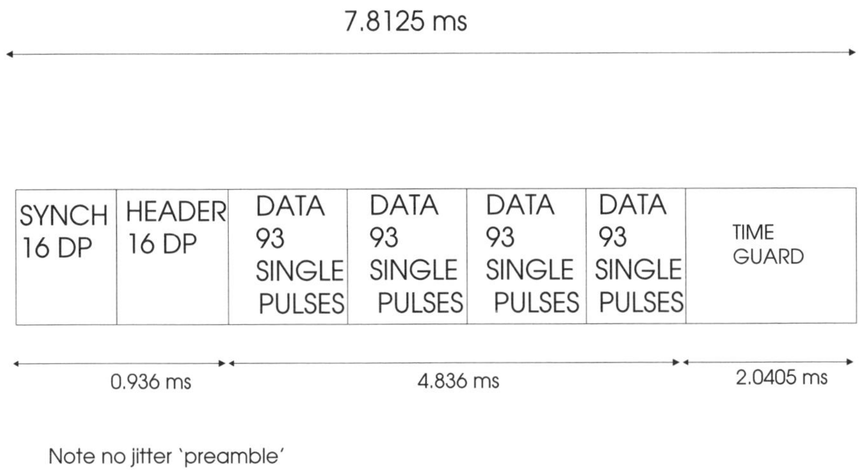

7.8125 ms

SYNCH 16 DP	HEADER 16 DP	DATA 93 SINGLE PULSES	DATA 93 SINGLE PULSES	DATA 93 SINGLE PULSES	DATA 93 SINGLE PULSES	TIME GUARD

0.936 ms   4.836 ms   2.0405 ms

Note no jitter 'preamble'

Figure 4.13 Packed 4 single pulse.

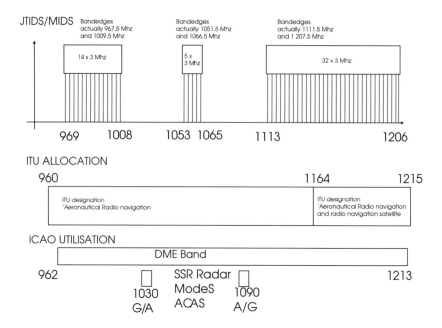

Figure 4.14   JTIDS/MIDS overlay with ITU band.

4.6.10   Other Salient Features of JTIDS/MIDS

The networks can be interfaced to IP networks or landlines.

There are different kinds of terminals LVT-1, LVT-2, LVT-3 and LVT-MOS, which, roughly speaking, correspond to equipment generation and application purpose.

There are inherent surveillance and navigation features. (These are not described in detail here.)

4.6.11   Overlay with DME Band

The above sections describe how the JTIDS/MIDS has been engineered to share the conventional bands described internationally in ITU (Figure 4.14). The basis of operation is on a non-interference basis; this is managed nationally usually via frequency clearance agreements between the National Military and the Radio Regulator or Civil Aviation Authority.

5 Long-Distance Mobile Communications

Summary

This chapter covers the various alternatives for achieving long-range communications. Long range in this context means 'over the horizon' communications. This can be achieved by using high-frequency (HF) radio systems or satellite radio systems. It can also be achieved by other means such as diffracted path radio systems with very high power or by using other networks such as cable or a hybrid of radio and landline. As such, for both the HF and satellite radio systems, some additional backhaul is generally required.

5.1 High-Frequency Radio – The Legacy

As was discussed already in the VHF section (Chapter 3), the very first aeronautical frequencies used for communications were actually HF. At the time this was for two reasons. Firstly the equipment was easy to make, with reasonably high-power amplifiers, and secondly the antenna systems were and still today are highly efficient so that most of the power being launched goes into the radio beam.

The HF band, as described in the Theory section (Chapter 2), lends itself to extensive range propagation with a low loss of reflection possible from the ionosphere, enabling various 'modes' to propagate thousands of kilometres and in some cases right the way around the world. This can be a blessing (i.e. when long range is the objective, sometimes thousands of kilometres over multiple international states, curving around a significant part of the earth's globe) when in remote parts of the world or in areas of poor VHF coverage, but it can also equally be a hindrance in that unwanted radio transmissions can equally propagate a long way. The overall effect can be a build-up or aggregation of unwanted signals, which looks like interference to the HF receiver and an apparent lift in the noise floor. In fact with HF this is generally the case.

Aeronautical Radio Communication Systems and Networks   D. Stacey
© 2008 John Wiley & Sons, Ltd

Also the availability of radio channels in the HF band is a function of which channels are 'open' and 'closed' at the time. This is ultimately dependent on time of day or night, sunspot activity, solar flares (which are to a degree random) and range.

As a consequence of these two aspects, aviation has been awarded spectrum allocations right across the HF band between 2.8 and 30.0 MHz to ensure a number of channels are open at any one time. In approximately 27 MHz of spectrum, civil aviation has been allocated 1.5 MHz in total, with the off-route (military function) being awarded a similar separate additional portion.

5.2 Allocation and Allotment

This allotment plan was established at WRC 1978 and is commonly known as Appendix 27 to the radio regulations. The allocation has not changed substantively since then. There is increasing pressure on the services in the HF band to be used more efficiently and to minimize spurious intrusion into adjacent channels.

To this end it was decided that after 1981, the use of the inefficient double-side band (DSB-AM) modulation would be phased out in favour of single-side band (SSB) modulation (conventionally the upper side band is used). This would immediately yield a double increase in channel capacity of the band as a whole. In 1995, the structure of the HF band was further redefined under ITU as Appendix S.27 to the radio regulations.

5.3 HF System Features

Figure 5.1 describes how aeronautical HF is synthesized, modulated and demodulated.

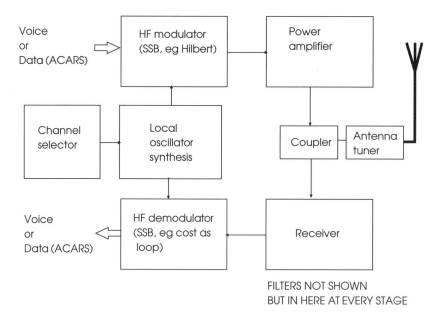

Figure 5.1   Aeronautical HF synthesis, modulation and demodulation.

5.3.1 Transmitter

The transmitter is synthesized using Carson's loop process (described in Chapter 2 Theory section) or by direct filtering out of the lower side band and carrier. In general, for voice modulation the voice is band limited between 300 and 2700 Hz. The reasons being, as previously discussed, that this is where the majority of useful voice power spectrum lies and it minimizes spill over of out of channel power in the RF stage. The transmitters are tunable in 1-kHz steps, which is in keeping with the ITU band plan allotment defined for aeronautical HF in Appendix S.27.

The RF emission has a given characteristic as a function of the pre-filtering, carrier filtering and post- (RF-) filtering. This is to ensure compatibility with co-channel and adjacent channel emissions. The emission masks have given characteristics; for the AM(R)S service in HF these can be J3E, J7B and J9B (for selective calling, sometimes called SELCAL, it is H2B); for more detailed discussion of these, see ITU-R definitions.

For aircraft stations the allowable peak envelope power is generally up to 26 dBW (400 W max into antenna Tx line) (in some special cases up to 600 W is allowed), and for ground stations, emissions Tx power up to 37.78 dBW (6 kW); these limits come from the ICAO SARPs and ITU Appendix S.27.

The frequency accuracy of airborne transmitters must be within 20 Hz of allocation; for ground transmitters this is even more stringent and must be within 10 Hz of allocation (3.33×10^{-5} %). This fairly tight specification is to safeguard out-of-band emissions into adjacent channels. With today's technology, this is easy to meet.

5.3.2 Receiver

The receiver deploys a Carson's loop receiver (described in Theory section). The frequency stability of the receiver function from the ICAO SARPs must be less than 45 Hz. On the surface, this may seem inconsistent with the transmitter specification; however, the capture effect will ensure the HF receiver is locked onto the received signal. A typical sensitivity of the receiver is quoted as 2 µV/m for 6 dB $S/(S + N)$ for the J3E-type emission.

5.3.3 System Configuration

The aeronautical HF system channels can be operated in two different ways.

Open channels shared between multiple users in broadcast mode enable a channel to be used in simplex.

Selective calling channels are used bidirectionally in half-duplex mode between an aeronautical station and a ground station. The two parties identify each other via a unique station number (much like a telephone DTMF Code).

5.3.4 Selective Calling (SELCAL)

To permit the SELCAL of individual aircraft over radiotelephone channels linking the ground station (or other aircraft) with the aircraft, the individual stations must be allocated a calling code. There are 16 300 codes available in the world.

SELCAL is accomplished by the coder of the ground transmitter sending a simple group of coded tone pulses to the aircraft receiver and decoder. It uses multifrequency dual tones exactly like the more modern telephone exchanges.

When using HF, there can be a small delay compared to the perceived 'instantaneous' nature of VHF. This is because the propagation time delay is a function of distance. So for a long path (say 9000 km), this can get up to as much as 30 ms. This is just noticeable but not as severe as satellite systems.

5.3.5 Channel Availability

When using an HF radio system the propagation conditions continually change with day or night, sunspot activity, solar flares, range and mode, etc. So it is necessary to first establish which channels are open and reliable.

Also sometimes it is possible that a channel can be lost whilst it is being used as these propagation conditions change; this is usually only the case for the longer distance connections involving skips or multiple skips. In this case the call could drop out and the users would likely transfer to another channel or go through the call set-up process again (Table 5.1).

ICAO has defined the world surface as split up into 14 separate zones or geographical areas that it calls 'major world air route areas' (MWARAs). Such areas are, for example, the North Atlantic, the Caribbean, the north Pacific, Southeast Asia, South America, Indian Ocean. (The detailed delineation can be found in ICAO Annex 10). Each one of these has its own HF ground station(s).

The ground stations will be allotted a number of frequency channels. These are a sub-band of the total aviation frequency pool. They are chosen to minimize interference into adjacent MWARAs and to give each MWARA a broad range of available channels across the total 27 MHz so that at any given time, some or most of the channels are open for reliable HF communications.

Table 5.2 describes the existing allotment of channels between the various MWARAs. For example, the North Atlantic can be seen to operate 24 channels right across the HF band.

Also some countries, particularly those with large geographic areas and large oceanic responsibility and relatively low populations that make it uneconomical to sustain a large VHF network (e.g. South Africa, Brazil and Australia), have a sizeable regional and domestic air route areas where HF is well deployed.

Emergency frequencies are allocated within this allocation on a regional basis (e.g. the United States; 3303 kHz) and also on an international basis (see Search and Rescue 3023, 5650 and 5680 kHz). There are also provisions for some specialist services, for example, Space Shuttle Recovery 5180, 5190 kHz (the space shuttle acts as an aircraft for its final stages of landing and therefore legitimately must be included in the AM(R)S service), as well as for coordinating flight tests worldwide (5451, 5469 kHz). In addition, for long-range weather information a Volmet (this is a meteorological service within a given volume) Service is broadcast. This can give meteorological conditions for zone, large area or even an individual airport. For example, Volmet North Atlantic broadcasts on 2905, 3485, 59, 6604, 8870, 10 051, 13 270, 13 276 kHz.

Table 5.1 Aeronautical HF allocation: civil and military (AM(R)S and AM(OR)S).

Frequency (MHz)	Channels	Designation	ITU footnotes	Permissible emissions
2.850–3.025	57 × 3 kHz	Aeronautical mobile (R) service	5.111, 5.115	A3E, H3E, J3E and H2B for SELCAL (also J7B and J9B as applicable)
3.025–3.155	43 × 3 kHz channels	Aeronautical mobile (OR) service		J3E and H2B for SELCAL (also J7B and J9B as applicable) + military
3.4–3.5	33 × 3 kHz channels	Aeronautical mobile (R) service		J3E and H2B for SELCAL (also J7B and J9B as applicable)
3.90–3.95	16 × 3 kHz channels	Aeronautical mobile (OR) service		J3E and H2B for SELCAL (also J7B and J9B as applicable) + military
4.65–4.70	16 × 3 kHz	Aeronautical mobile (R) service		J3E and H2B for SELCAL (also J7B and J9B as applicable)
4.70–4.75	16 × 3 kHz	Aeronautical mobile (OR) service		J3E and H2B for SELCAL (also J7B and J9B as applicable) + military
5.45–5.48	9 × 3 kHz	Shared fixed Aeronautical mobile, Aeronautical mobile (R) and (OR) service + land mobile (region variations)		J3E and H2B for SELCAL (also J7B and J9B as applicable) + military
5.48–5.68	66 × 3 kHz	Aeronautical mobile (R) service	5.111, 5.115	A3E, H3E, J3E and H2B for SELCAL (also J7B and J9B as applicable)
5.68–5.73	16 × 3 kHz	Aeronautical mobile (OR) service		J3E and H2B for SELCAL (also J7B and J9B as applicable) + military
8.815–8.965	150 × 3 kHz	Aeronautical mobile (R) service		J3E and H2B for SELCAL (also J7B and J9B as applicable)
8.965–9.040	13 × 3 kHz	Aeronautical mobile (OR) service		J3E and H2B for SELCAL (also J7B and J9B as applicable) + military
10.005–10.100	16 × 3 kHz	Aeronautical mobile (R) service	5.111	J3E and H2B for SELCAL (also J7B and J9B as applicable)
11.175–11.275	66 × 3 kHz	Aeronautical mobile (OR) service		J3E and H2B for SELCAL (also J7B and J9B as applicable) + military

(Continued)

Table 5.1   (*Continued*)

Frequency (MHz)	Channels	Designation	ITU footnotes	Permissible emissions
11.275–11.400	42 × 3 kHz	Aeronautical mobile (R) service		J3E and H2B for SELCAL (also J7B and J9B as applicable)
13.20–13.26	20 × 3 kHz	Aeronautical mobile (OR) service		J3E and H2B for SELCAL (also J7B and J9B as applicable) + military
13.26–13.36	33 × 3 kHz	Aeronautical mobile (R) service		J3E and H2B for SELCAL (also J7B and J9B as applicable)
15.01–15.10	33 × 3 kHz	Aeronautical mobile (OR) service		J3E and H2B for SELCAL (also J7B and J9B as applicable) + military
17.90–17.97	13 × 3 kHz	Aeronautical mobile (R) service		J3E and H2B for SELCAL (also J7B and J9B as applicable)
17.97–18.03	20 × 3 kHz	Aeronautical mobile (OR) service		J3E and H2B for SELCAL (also J7B and J9B as applicable) + military
21.924–22.000	25 × 3 kHz	Aeronautical mobile (R) service		J3E and H2B for SELCAL (also J7B and J9B as applicable)
23.200–23.350	50 × 3 kHz	Aeronautical mobile (OR) service		J3E and H2B for SELCAL (also J7B and J9B as applicable) + military
Total	(480 (R)) (352 (OR))			

5.4   HF Datalink System

In parallel to technological advances and the development of a VHF datalink, there was the same requirement to pass data between aircraft and the ground (or other aircraft) for new applications (for improving the navigation, surveillance and communication functions) for long-range flights. Thus some kind of HF datalink was required. Unlike the ACARS over VHF, the ACARS for HF has followed the ICAO standardization process, and in 1999, HF ACARS was incorporated in Annex 10 (Amendment 74). Provisions for this new HF datalink service were approved by the ITU in July 1998.[1]

5.4.1   Protocol

ARINC standard 635-2 defines a data protocol, modulation, data interface and specification for operating HF datalink (Figure 5.2). It uses a relatively primitive protocol using PSK modulation at a gross bit rate of 1800 bps.

Table 5.2 AM(R)S parameters

(a) Evolution

Satellite	System			
Generation	1*	2	3	4
Capacity channels	50	250	1500	25 000+
Power (EIRP)	32–35 dBW	39 dBW	48 dBW	60+ dBW
Features	GEO, global	GEO, global	GEO, global and	GEO + possible
Commercing	beam, leased	beam	spot beams	lower earth orbits
Year of operation	1982	1992	1995	1998

(b) Inmarsat 2 services

Mobile	Earth				
Parameter	Inmarsat-A	Inmarsat-B	Inmarsat-C	Inmarsat-M	Aeronautical
Steering	Steerable	Steerable		Steerable AZ only	Electronic
Type	Parabolic	Parabolic	Omnidirectional	Linear array	Phased array
EIRP	36 dBW	25–33 dBW	12 dBW min at 5°	22–28 dBW	14 dbW l-g
					26 dBW h-g
Receive G/T	−4 dB/K min	−4 dB/K	−23 dB/K at 5°	−12 dB/K	−26 dBK l-g
					−13 dBK h-g
Telex & data	50 baud telex	50 baud telex	600 bps	2.4 kbps	600, 1200, 2400,
rates		9.6 kbps			10 500 bps
Telephony	FM	Digital coded	Data only	Digital coded	Digital coded
Voice coding	12 kHz	16 kbps		4.2 kbps	9.6 kbps
rates	deviation				
Channel	50 kHz	20 kHz	5 kHz	10 kHz	5 kHz
spacing					17.5 kHz

5.4.2 Deployment

Data base stations are deployed at most of the MWARA sites worldwide. ARINC provides a service through these (its 'GLOBALink' service connecting 13 sites), which offers full coverage of the globe, north of the −70° latitude mark. (There is very little civil aviation flight activity below this line, other than transcontinental flights from Australasia to South America (which are only a few per week).)

The channels allotted to HF ACARS frequencies include 3007, 6646, 6712, 8942, 8977, 10 084, 11 384, 13 339, 15 025, 1799 kHz.

5.5 Applications of Aeronautical HF

Civil and military aviation mainly use HF for long-distance en route over water or remote regions where no VHF coverage exists. In some instances it can act as an emergency backup to the VHF system (or the satellite communications system). It is cheap to operate as ground infrastructure is minimal for the vast areas of coverage provided. For oceanic flights, carriage

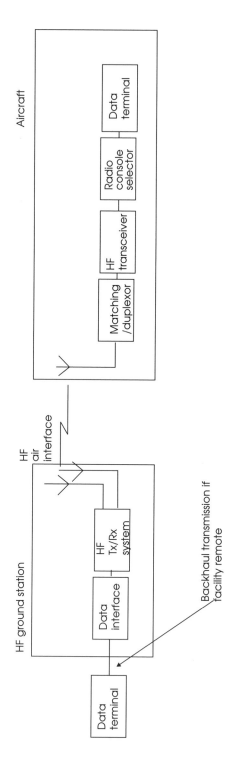

Figure 5.2 HF ACARS system block diagram.

of HF avionics is usually mandated by the regulatory authorities involved (the Civil Aviation Authority for the region or the Joint Aviation Authority, JAA).

HF systems tend to run a conservative link budget. Availability of a channel is rarely due to it not having enough power to reach through but is more often a function of a channel being open or closed at a given time and adverse and random propagation conditions. Also with the continuing rise of interference, the $S/N + I$ (which approximates S/I when interference is dominant) becomes a bigger issue than link budget. One example of this is in the south Pacific where there is significant 'pirate' activity in both senses of the word (illegal vessels and illegal frequency operation), and it is often found that maritime vessels use aeronautical HF channels to broadcast their messages.

The semi-random element to availability (reliability, integrity) makes HF engineering some-times more of a black art than a science. Different countries and airlines use or do not use HF to different extents. There are now sophisticated automated systems available to minimize pilot/user interface. They can take the pain out of establishing a link. They use a combination of sophisticated software and ionosphere sounding techniques to optimize this.

5.6 Mobile Satellite Communications

5.6.1 Introduction

5.6.1.1 Geostationary Satellite Systems

To date, civil aviation mainly uses satellite services from geostationary satellites. That is, satellites in a stationary orbit above the earth's equator at a distance from the earth's sur-face of 35 790 km or 42 160 km from the centre point of the earth's sphere. These satellites seem stationary with regard to the position of the earth. In particular, aviation widely uses the Inmarsat services for safety and regularity of flight applications. The service is called the aeronautical mobile satellite (route) service (AMS(R)S) and is the satellite equivalent of the AM(R)S (Figure 5.3). (The off-route equivalent as used by the military and governments is referred to as the aeronautical mobile satellite (off-route) service or AMS(OR)S.)

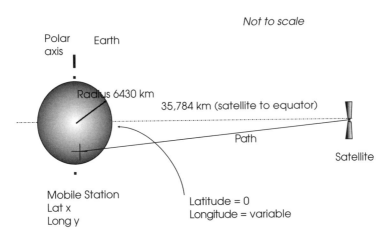

Figure 5.3 Geostationary satellite systems.

Table 5.3 Inmarsat ground stations.

Satellite (location)	Atlantic East Europe Inmarsat 3 F2 (15.5° W)	Pacific Ocean Inmarsat 3 F3 (178° E)	Indian Ocean Inmarsat 3 F1 (64° E)	Atlantic West 3 F4 (54° W) Ocean W
Inmarsat ground station	Goonhilly, Eik, Aussaguel	Yamaguchi, Sentosa, Perth, Santa Paula	Yamaguchi, Sentosa, Perth, Aussaguel	Goonhilly, Eik, Santa Paula

These satellite services have multiple allocations in L band (mainly around 1.4 GHz) for the mobile portion, C Band (3.4–4.8 GHz), Ku (10.7–12.75 GHz) and Ka (19–22 GHz) bands mainly for the fixed portion. Many military systems exist and these are not fully disclosed to the public domain. They will not be discussed further in this book; however, the systems described in the civil satellite applications are usually duplicated in the military domain.

5.6.1.1.1 The Inmarsat System

For the Inmarsat system, there are a number of ground stations distributed throughout the world (Table 5.3). Usually there are at least three for each oceanic region; for example, in Europe there is Goonhilly in the UK, Eik in Norway and Aussaguel in France – all acting as ground station (or hubbing points) for the Inmarsat aero service to the Inmarsat 3-F2 satellite at 15.5° west.

These form the trunking points for the satellite mobile network. This provides safety of life and non-safety of life type services (Figures 5.4 and 5.5).

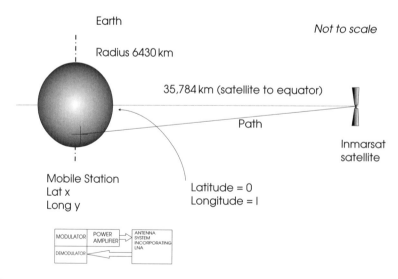

Figure 5.4 Inmarsat mobile system components.

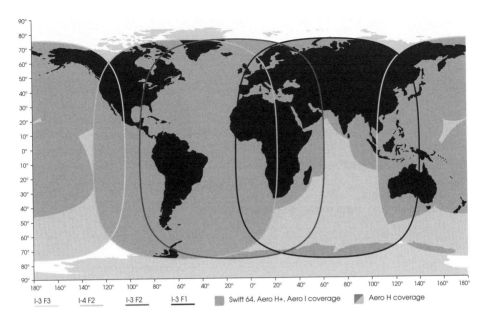

| I-3 F3 | I-4 F2 | I-3 F2 | I-3 F1 | Swift 64, Aero H+, Aero I coverage | Aero H coverage |

Figure 5.5   Inmarsat mobile system coverage. Reproduced by permission of Inmarsat.

From the satellite hub points voice and data is back hauled back over the conventional public service telephone network (PSTN).

Passenger and Non-Safety Services.   For aircraft passenger communications, which is considered generally non-safety, and regularity of flight, the aeronautical mobile satellite service (AMSS) is used. There are service providers in the extended Ku band (14.5 GHz), again based on geostationary satellite orbit (GSO) satellites, providing these commercial services to aviation over geostationary satellites.

5.6.1.2   Low-Earth Orbit Satellite Systems

To recap from the Theory section: Generally speaking, low-earth orbit satellite systems or LEOs can be envisaged to operate between 600 and 2000 km above the earth's surface. (However, the definition of LEO is said to vary; for example, some definitions even include LEOs up to the geostationary distance at 35 790 km.) This gives a round trip delay of between 4 and 70 ms.

Currently aviation does not use LEO satellites to provide AMS(R)S. However, this shouldn't necessarily be precluded from the future. It is possible to use the AMS service provisions onboard aircraft that have been set up with the necessary receivers (Figure 5.6). Such services are offered by Iridium/Globalstar. The main advantage of using this service is the low time delays for voice and the low terminal size with omnidirectional antennas and relatively high data bandwidths.

The up and down mobile links use 1.4 and 2.4 GHz frequencies. The feeder links providing backhaul to the earth station operates above 5 GHz and is shared to a degree with the aeronautical radio navigation (microwave landing system band). The link budget for a LEO system is favourable compared to other satellite systems due to the relatively lower free space path loss.

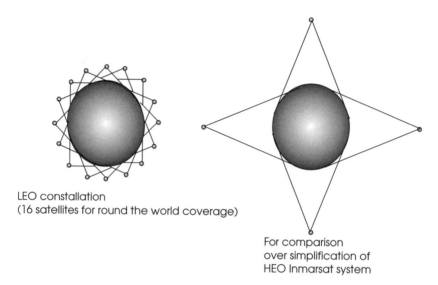

LEO constellation
(16 satellites for round the world coverage)

For comparison
over simplification of
HEO Inmarsat system

Figure 5.6   LEO satellite system.

5.6.1.3   Medium-Earth Orbit Satellite System

The range of orbits of a medium earth orbit satellite is not well defined but generally they can be considered to operate between 3000 and 30 000 km. This gives a round trip delay of typically between 20 and 250 ms. No service providers are known to provide a general service to the aeronautical civil application and although not discounted in the future, it is not discussed further.

5.6.2   Geostationary Services System Detail

5.6.2.1   The AMS(R)S Satellite System

Considering in detail the AMS(R)S specification laid down by ICAO in Annex 10 of the ICAO convention,[2] in general, global (broad) beams are required to give wide coverage over the earth's surface; these are particularly required in oceanic regions and in low-population regions where VHF coverage is uneconomic. Typically, a few global beams (three to be precise) will give almost total coverage of the world's surface (and four comfortably with overlap of the key population are as required by Inmarsat) except in some of the extreme polar regions which will always have the poorest coverage; this is due to the geometry (see Theory section) of geostationary satellites, giving a horizontal look angle from the poles and in addition the longest path of radio propagation (in terms of free space path loss this is an extra 7000 km to travel or an extra 2 dB of attenuation, which on a tight link budget can be hard).

Additional spot beams are included to provide capacity for high-activity areas (e.g. Europe). These beams encompass significantly less than the earth's surface in LOS, 'usually designed to improve system resource efficiency with respect to geographical distribution'.

The round trip propagation delay varies from 240 to 280 ms. Call set-up times are consequentially typically a few seconds as the air interface goes through a handshake. Due to the

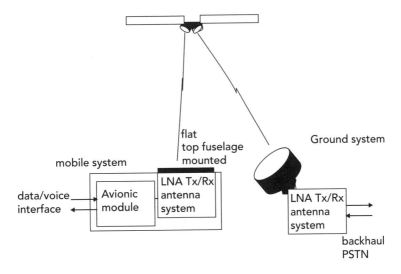

Figure 5.7 AMS(R)S system.

above constraints, i.e. voice and call set-up delays, geostationary satellite communication is not instantaneously available and therefore is not likely to be suitable for time critical functions as required in the take-off and final stage of landing operations. It is, however, ideal for en route communications, particularly when an aircraft is up above the precipitation layer and particularly on a long-range route where it could be in coverage from one (or at most two) satellite for the whole duration.

The links are high reliability (each satellite is duplicated in space with standby satellites available) (Figure 5.7).

5.6.2.1.1 Modulation

For the lower bit rate service provisions, this is specified to be aviation binary shift keying (A-BPSK) running at a gross system bit rate of 2.4, 1.2 or 0.6 kbps. For higher bit rates of greater than 2400 kbps, aeronautical quadrature shift keying (A-QPSK) is deployed. The satellite is controlled from ground by a network coordination station.

5.6.2.1.2 Protocol

The AMS(R)S protocol follows the OSI (open system interconnection) layered model (Figure 5.8). Call set-up is facilitated by the P and R channels. The data payload is taken on the T and C channels.

The P channel operates in packet mode, using time division multiplexed frames transmitted continually from aeronautical ground earth station to the aircraft via the transponder on the satellite; this provides the signalling data for the link set-up and synchronization.
The R channel is a random access protocol using a slotted Aloha format. This is transmitted from the aircraft (or mobile station) when initiating a call or data message set-up. It can also carry local synchronization and network management information.

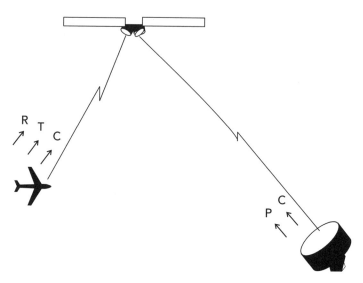

Figure 5.8   AMS(R)S system connection.

The T channel is a data channel using the TDMA protocol for assignment of slots channel from aircraft only. Receiving geostationary earth station (GES) reserves time slots according to message length.

The C channel(s), circuit mode channel, single channel per carrier are used both ways.

For a fuller detailed explanation of the protocols, see ICAO Annex 10, Volume III, Part I, Chapter 4.

5.6.2.1.3 Call Set-Up Delay

In Annex 10, ICAO specifies what should be the maximum call set-up times. These may seem rather long but bear in mind the handshake required to perform the call set-up over the P, R, T and C channels involves multiple messages being sent each way serially over the satellite transponder (Table 5.4). In reality of course calls are much faster than this.

Table 5.4   Call set-up times.

Minimum channel speed (bps)	Maximum transmit delay high priority	Maximum transmit delay low priority
600	12	40
1200	8	25
2400	5	12
4800	4	7
10 500	4	5

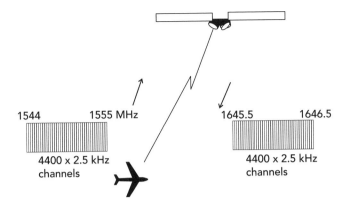

Figure 5.9   Aeronautical mobile satellite channel plan.

5.6.2.1.4   Priority and Pre-Emption

AMS(R)S shall have priority over all non-AMS(R)S calls and shall be capable of pre-empting non-AMS(R)S calls. This is an agreement made between ICAO and the satellite providers. It has an interesting legacy. A decade ago, the AMS(R)S system had exclusivity in the L band where it was allocated. Due to pressure from other mobile satellite users, it was decided at WRC 2000 to remove this, however, to put a priority to the AMS(R)S allocation being provided. In practice it is very hard to see if this is going to be honoured, and the mechanism for ensuring these provisions is still to be demonstrated by the satellite provider.

5.6.2.1.5   Allocation and Allotment

Transmitter bands (Figure 5.9):

- To aircraft 1544–1555 MHz (distress and safety 1544–1545);
- From aircraft 1645.5–1646.5 MHz;
- Tuning in 2.5 kHz increments;
- Channel number $C_t = (f_{Tx} - 1510.10)/0.0025$ (approx. 4400 channels altogether);
- Aircraft receiver can deal with Doppler change of 30 Hz/s.

5.6.2.1.6   Grade of Service

Probability of blocking when setting up an AMS(R)S call is defined as being less than 1 %.
The round trip transfer delay: the propagation plus electronics/network transfer has to be
   <0.485 s.
Probability of misrouting by internal processing error or signal error by GES $<1 \times 10^{-6}$ (see
   CRC codes discussion in the Theory section).

5.6.3   Antenna System Specifications

The aeronautical mobile antenna (mounted on the aircraft normally on the upper side to give best LOS conditions) specification is as follows (Figure 5.10):

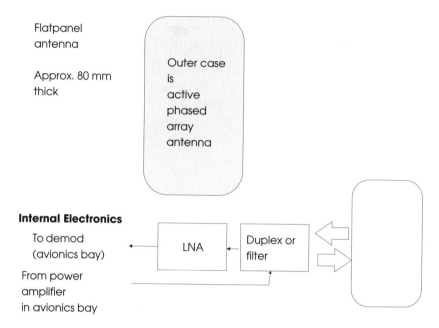

Figure 5.10　Aeronautical mobile satellite antenna specification.

Note the antenna system (includes the antenna and the first stage of the low-noise amplifier), which is characterized by the *G/T* figure or 'figure of merit' as discussed in the Theory section.

The system has to have omnidirectional* coverage 360° for all aircraft aspects (elevation 5–90°; for aircraft attitudes of +20° to −50° and for aircraft roll of ±25°).

The polarization is right-hand circular.[3]

5.6.3.1　Satellite Antenna Figure of Merit (G/T)

Two types of antenna system are defined.

- Low-gain antenna subsystem $G/T > -26$ dB/K over 85 % of reference coverage zone. $G/T > -31$ dB/K over remaining 15 % of volume.
- High-gain antenna subsystems $G/T > -13$ dB/K over 75 % of reference coverage zone $G/T > -25$ dB/K over remaining 25 % of volume.

5.6.3.2　Antenna Discrimination

- Antennas must discriminate > 13 dB between direction of wanted and unwanted satellite spaced 45° or greater in longitude over 100 % of coverage area. This is to increase the spectral efficiency of a system and increase the reuse potential in one of the most congested/utilized bands in all the radio spectrum.

---

* Note that the antenna does not necessarily have to be a passive omnidirectional, but can be an active or mechanical directional antenna able to steer beam, and in reality this is the case.

5.6.3.3 Rx Thresholds

These are defined as following:

- P-channel average bit error rate (BER) $<10^{-5}$ when illuminated by power flux density (PFD) $= -100\,\mathrm{dBW/m^2}$ on central lobe;
- C-channel average BER $<10^{-3}$ when illuminated by PFD $= -100\,\mathrm{dBW/m^2}$ on central lobe.

Table 5.5 Comparison of voice provision over different systems.

	VHF	L-band communication (JTIDS, MIDS)	HF communication	Satellite
Frequency	118–137 MHz	960–1215 MHz	3–22 MHz	1544–1646.5 MHz
Channelization	100, 50, 25, 8.333 kHz, up to 2100 channels	3 MHz × 51	3 kHz × 600+	(approx. 10 MHz up + 10 MHz down) 2.5 kHz approx. 4400 channels
Type	Simplex	Duplex	Simplex	Duplex
Modulation type	Analogue and digital DSB-AM (legacy) TDMA (VDL3) 8DPSK (VDL3)	Digital FDD, TDMA, CDMA, frequency hopping	SSB	
Path reliability	Good at LOS, lower can be unpredictable, so ionospheric refection can cause problems	Not declared	Medium can be unreliable	Very good
System availability	>99.9–99.999 %		System >95–99 % channels much less	99 %+
Propagation delay	<1 ms	<1 ms	<10 ms	Typically 250 ms
Voice bandwidth	Can be 4.8 kbps (in VDL3 codec, Nyquist rate of 0–2400 Hz)	32 kbps	<3 kHz	4.8 kbps

Table 5.6 Comparison of data provision over different systems.

	VHF	L-band communication (JTIDS, MIDS)	HF communication	Satellite
Frequency	134–137 MHz (and 108–118 for VDL4 only)	960–1215 MHz	3–22 MHz	1544–1646.5 MHz
Channelization	25 kHz VDL	3 MHz × 51	3 kHz × 600+	(approx. 10 MHz up + 10 MHz down) 2.5 kHz approx. 4400 channels
Type	Simplex (VDL3Duplex)	Duplex	Duplex	Duplex
Modulation type	Digital CSMA, TDMA, 8DPSK, GFSK	Digital FDD, TDMA, CDMA, frequency hopping	Digital 8PSK	Digital A-BPSK Modern QPSK
Path reliability	Under study	Extremely high	Medium can be unreliable	Very good
System availability	> Under study (objective 99.9–99.999 %)	Not openly defined	System > 95–99 % channels much less	99 %+
Data throughput (gross)	VDL 2 19.2 kbps VDL 3 19.2 kbps VDL 4			
Propagation delay	Sub ms	Sub ms	<10 ms	Typically 250 ms

5.6.3.4   Tx EIRP Limits

- Low-gain antenna EIRP >13.5 dBW in direction of satellite EIRP <22.8 dBW in any direction.
- Intermediate gain antenna EIRP >12.5 dBW in direction of satellite EIRP <34.8 dBW in any direction.
- High-gain antenna EIRP >25.5 dBW in direction of satellite EIRP <34.8 dBW in any direction.
- At non-maximum setting EIRP shall be backed off 5 dB on level towards satellite carrier.
- Off-level EIRP should be −24.5 dBW in any direction.
- Intermodulations shall not cause problems to sat nav receiver (fifth order intermodulation below 1610).
- Simultaneous operation of C&R or C&T channels allowed but not all three (C, R and T).

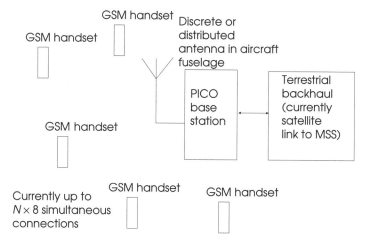

Figure 5.11   System diagram aeronautical passenger communications. (Planned.)

This specification sets some interesting engineering boundaries, mainly to maximize utilization of the scarce L-band spectrum and maintain compatibility between adjacent aeronautical mobile transceivers.

5.7   Comparison Between VHF, HF, L Band (JTIDS/MIDS) and Satellite Mobile Communications

As a final review in this chapter, it is worth comparing the different qualities of the four main mobile radio systems on-board a typical aircraft at a system level. Tables 5.5 and 5.6 show this and their relative strengths and weaknesses.

5.8   Aeronautical Passenger Communications

There are a number of systems for airline passenger communications, both for voice and data. They use satellite backhaul that operates in Ku band (14.5 Hz). Figure 5.11 illustrates a current plan to enable the use of mobiles and data terminals on-board aircraft using a satellite backhaul. A previous system operated by Connexion by Boeing also used a similar infrastructure but was only able to handle data.

Further Reading

1. ICAO standardization (ref RR 27/15, 27/19, 27/58)
2. ICAO Annex 10, Volume III, Part I, Chapter 4
3. See ITU SI.154

6 Aeronautical Telemetry Systems

Summary

This chapter provides a description of some of the aeronautical telemetry systems deployed today. In addition, currently intensive work is underway defining and building new aeronautical telemetry systems using state-of-the-art technology and enabling higher bit rates. These are described in greater depth. Spectrum allocation for the new systems is expected in a number of bands in ITU WRC 2007, and the system definitions for this next phase are mature enough to be explained here. Finally, unmanned aerial vehicles (UAVs) that are already used today in limited and non-controlled airspace are expected to see a growth over the next decades.

6.1 Introduction – The Legacy

Back in the 1950s and 1960s when aircraft development and manufacture was arguably going through a revolution, as world demand for civil flying started to rise, there was a consequential growing need for aircraft testing. It was soon realized that it was preferable during the aircraft testing cycle to bring down aircraft dynamic parameters as they changed during these flight tests because it enabled the relatively dangerous test flight durations to be reduced. For this, telemetry was developed. Telemetry was mainly (sometimes totally), a one-way link between the aircraft and the ground for downlinking (i.e. aircraft to ground) aircraft speeds, aspects, altitudes as well as for stress and strain parameters of the airframe and some of the component parts such as engine parameters, subsystems such as the fuel systems right through to avionics (Figure 6.1). It is well accepted, by bringing down as much information as possible, that flight tests could be minimized. This indirectly minimized safety risk and risk to test pilot lives at the time, and is arguably still true today.

The limitation with the first telemetry systems was not the radio communication bearer but the mainframe computer technology, which did not change till the 1970s. Telemetry systems are usually used on a test range in restricted airspace, which has a geographic locality. (For example Airbus industries has traditionally used an area of about 200 km around Toulouse in

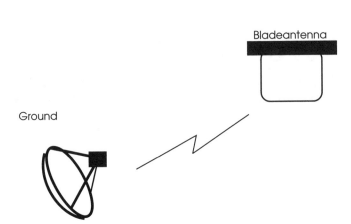

Figure 6.1   Aeronautical telemetry link.

France with additional extensive testing going on in dedicated airspace over the bay of Biscay where risk to other aircraft and the public can be minimized as prototypes are flown.) By contrast, in the Unites States there are about 50 different 'ranges' distributed roughly equally across the country.

6.2 Existing Systems

Today, there are four main bands allocated to telemetry below 3 GHz, namely the 1435–1525 MHz and the 2200–2290 MHz bands used extensively internationally, the 2300–2400 MHz band used extensively in Europe and the 2360–2390 MHz band used in the United States. Above 3 GHz, the main bands are 4.4–4.99 GHz and 14.7145–15.1365 GHz (Table 6.1).

These bands are being used by commercial companies such as Airbus, Boeing and Bombardier, and also by the military divisions of these and other companies to test military aircraft (Figure 6.2). Allotment of these channels varies from state to state and can be dynamic from day to day. Typically in S band 2.3–2.4 GHz, this is channelized into 500 kHz slots (Figures 6.3 and 6.4).

Table 6.1   Existing telemetry spectrum.

Band (GHz)	Available spectrum (MHz)
1.435–1.525	90
2.20–2.29	90
2.3–2.4 (Europe only)	100
2.36–2.39 (North America sub-band)	30
4.40–4.99	599
14.7145–15.1365	422

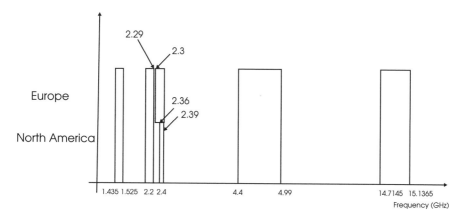

Figure 6.2 Telemetry spectrum.

Aircraft usually deploy an omnidirectional antenna system for all-round coverage and best flexibility. On the ground a tracking antenna is set up to follow the aircraft in its test flight path. This enables a better (longer range) link budget and therefore greater throughputs. It is, however not so spectrally efficient as the antenna is omni directional, broadcasting in all directions. Sometimes repeater stations, ships or other aircraft can be used to relay information and effectively enlarge the area of coverage.

Typically there maybe a number of aircraft under test (up to six aircraft is considered typical) on any one test range, and some of these could be autonomously piloted. Unmanned aerial vehicles or UAVs also offer the ability to test aircraft without even having the pilot on-board. In restricted airspace this is, maybe, the ultimate in air safety to public and staff.

6.2.1 A Typical Telemetry System Layout

Figure 6.5 shows a typical telemetry system configuration that would work with the single transmitter up to a distance of about 100 km (line of sight); with the optional booster amplifier,

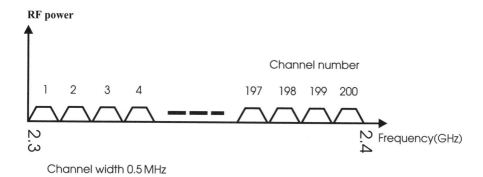

Figure 6.3 S band telemetry spectrum allotment (Europe).

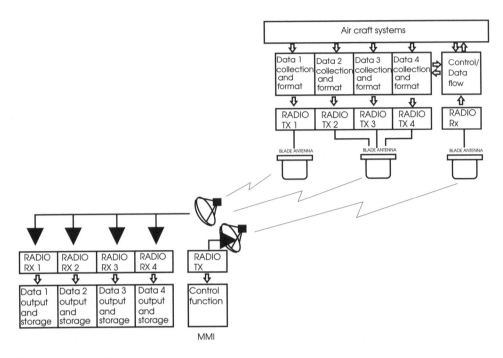

Figure 6.4   Typical radio system.

this range could be extended. The system is drawn for S band; however, equally, one of the other telemetry bands could be used in L band (1.4 GHz) or higher frequency band (14 GHz).

6.2.1.1   *Transmitter Side (On-board Aircraft Components)*

6.2.1.1.1   Baseband Inputs

A number of data feeds and digitized video are aggregated in a multiplexer. Typical data rates may be a couple of 64 kbps circuit's data and a video feed of 384 kbps in standard pulse code

Figure 6.5   Telemetry system in the S band 2.3–2.4 GHz.

modulation (PCM) format. Compression of the various signals can also be accomplished at this stage as necessary.

6.2.1.1.2 Multiplexer Output

The speed of the output of this stage can be typically 2 Mbps or in some equipment upto15 Mbps. An analogue video output can also be used.

6.2.1.1.3 S Band Transmitter

In the S band, data is modulated onto a straight frequency shifty keying (FSK) carrier. For a pure video system it can be modulated onto an frequency modulation (FM) analogue carrier. The RF bandwidth is 500 kHz.

6.2.1.1.4 Optional Power Amplifier

This stage can be added if the flight test range is going to be beyond typically 100 km or if a higher link reliability is required. This gives an optional 20-W power into the transmit antenna.

6.2.1.1.5 Omnidirectional Antenna

This is a low-gain, typically 2–3 dBi 'blade' antenna, usually mounted underneath the aircraft fuselage.

6.2.1.2 Receiver Side (High-performance Ground Station)

6.2.1.2.1 Tracking Antenna

A high-gain, typically >30 dB, narrow beam (less than a degree) parabolic dish antenna is used to track the aircraft. This is done using an elaborate tracking control system, which usually works with a feedback circuit, tracking for maximum receiver gain. However, variants have been seen using GPS mapping software or separate Tx/Rx paths for the antenna control circuits.

6.2.1.2.2 Low-Noise Amplifier

As is common good practice in receiver design, the receiver input is fed straight to a low-noise amplifier, with a typical gain of 30–40 dB with a typical noise temperature of 40 K.

6.2.1.2.3 Receiver Demodulator

Here the modulated FSK or FM signal is converted back into the digital PCM stream or analogue video format (such as NTSC or PAL).

6.2.1.2.4 Demultiplexor

Here the digital signal is demultiplexed into component parts.

6.2.1.3 On-board System Duplication and Ground Backhaul Infrastructure

Sometimes the on-board system can be duplicated for resilience of systems between the aircraft and ground, or where required bandwidths exceed the single channel capacity (for multiples of 2-Mbps or up to 15-Mbps bandwidth). These can be coupled onto one antenna or spread over two or more antennas, to give better link reliability, for example, when one of the antennas is shielded by the aircraft body.

Similarly the ground infrastructure can be duplicated, or a second ground station can be deployed remotely from the main telemetry collection centre and relayed back typically over fibre or radio links. There is also potential for an increasing migration towards integration with the Internet and IP networks for some of the distributed telemetry collection systems.

6.2.2 Telecontrol

Telecontrol is usually reverse to telemetry, i.e. transmission from ground to air. This can be used to control equipment on the aircraft and even the vehicle itself (a good example would be on UAVs). The comparative data rate and consequently bandwidth allocated for these functions is usually much smaller (the fraction is typically less than 5 % of the telemetry requirement). Usually the telecontrol part is carried out in another frequency band, either a telemetry band where radio regulations permit or in some countries even the AM(R)S band 118–137 MHz.

6.3   Productivity and Applications

As with most technology, the reason for a telemetry system is about increasing the productivity of an individual aircraft test so that expensive test flight programmes can be optimized and even shortened. Commercially there is a strong drive for this. In parallel to this, data communications and mobile communications have gone through technical revolutions. That is, today the processor bandwidth is ever spiralling up, and applications are becoming increasingly bandwidth hungry.

Both of these aspects have shown that today the scenario has reversed and the limitation for telemetry operations is the bandwidth pipe, i.e. the radio channel between the aircraft and the ground. For this reason there are numerous proposals at an international level (ITU) to expand the spectrum available for telemetry to enable higher capacity second-generation telemetry systems.

Investigations are going on into how the following bands can be further opened up to accommodate telemetry functions internationally. (Some of these bands are already operated on a national basis) (Table 6.2).

Table 6.2   Potential future telemetry bands.

Band	Available bandwidth (MHz)
4.400–4.940 GHz	540
5.030–5.091 GHz	61
5.091–5.150 GHz	59
5.150–5.250 GHz	100
5.925–6.700 GHz	775
22.5–23.6 GHz	1100
24.75–25.5 GHz	750
27.0–27.5 GHz	500

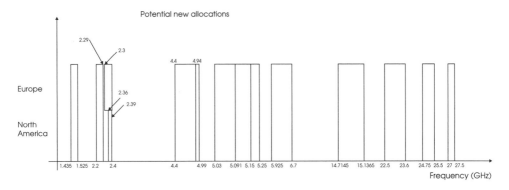

Figure 6.6   Proposed spectrum allocation.

6.4   Proposed Airbus Future Telemetry System

In Europe, to alleviate the telemetry capacity problem, Airbus is developing a new telemetry system around 5 GHz.

6.4.1   Channelization Plan

At the time of writing, Airbus is planning to deploy a new telemetry system that can accommodate up to five aircraft simultaneously in any one test range (the significant one being regional to the Toulouse area, with smaller ranges planned in the United Kingdom, Spain, Germany and Italy) (Figure 6.6).

Initially this will require up to five RF channels of notional 12 MHz bandwidth. This is planned between 5091 and 5150 MHz, but could be expanded above or below this band, subject to proving compatibility with radio local area network services, fixed satellite service feeder links in this upper band, and ARNS services of aviation below.

6.4.2   System Components

These are shown in Figure 6.7.

6.4.3   Telemetry Downlink

The system architecture is like the generic architecture described earlier just translated to the 5 GHz band. In fact the higher frequency increase in path attenuation is roughly compensated by the gains in ground station antenna. The proposed modulation is a form of coded orthogonal frequency division multiplexing (COFDM) (see Chapter 2), with individual sub-carriers spread with codes (Figure 6.8).

The airborne transmitters are planned to operate at about 20 W into an omnidirectional antenna of approximately 3-dBi gain. On the ground the tracking antenna will have a gain of >40 dBi and a beam width of less than a degree. The low-noise amplifier will have a gain of

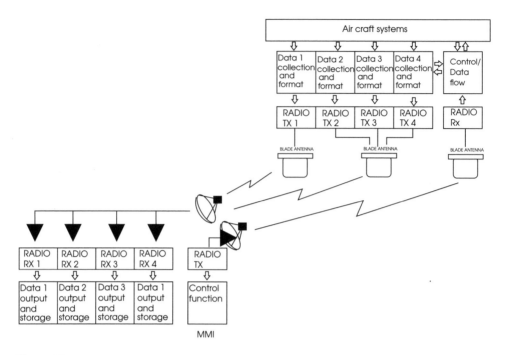

Figure 6.7   System block diagram.

>30 dB with a noise temperature of 40 K. The ground station receiver threshold will be below 100 dBm. The optimized data throughput for each telemetry channel will be up to 20 Mbps.

6.4.4   Telecommand Uplink

This is planned for the VHF communication band at 118–137 MHz, probably using >75 kHz (>3 × 25 kHz channels simultaneously). The transmit power will be >20 W at a highly directional tracking Yagi antenna. On the aircraft the conventional VHF communication blade (approx. 2–3-dBi gain) will receive this signal, and the required receive power at the antenna

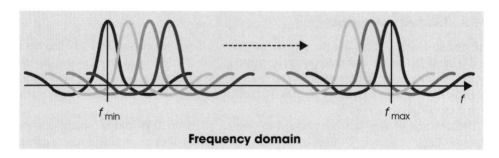

Figure 6.8   COFDM spectrum.

Top, tail-mounted omnidirectional antenna

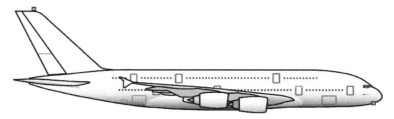

Bottom, front-mounted omnidirectional antenna

Figure 6.9 Aircraft antenna positions. Reproduced from Airbus Website.

must be approximately –90 dBm. The effective baseband throughput will be up to 150 kbps (Figure 6.9).

6.5 Unmanned Aerial Vehicles

In parallel to the planned future developments for telemetry, the market for UAVs is growing with some expectations showing them taking up a significant share of the commercial and military markets. For this telemetry and telecontrol spectrum is required. Allocations and allotments for this are expected to be further developed at WRC 2007 and WRC 2010 (Figure 6.10).

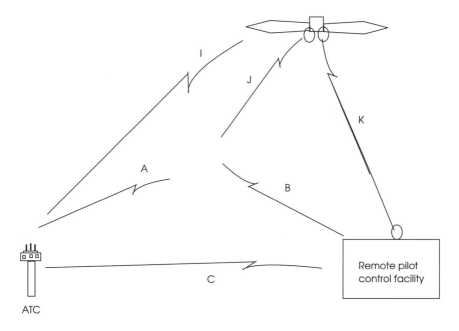

Figure 6.10 Unmanned aerial vehicle projected requirements.

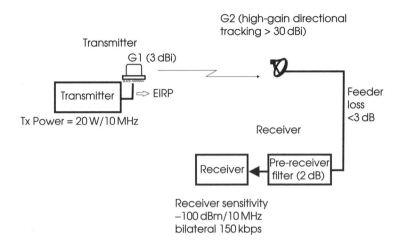

Figure 6.11   UAV hypothetical system block diagram. (Numerous RF frequencies proposed.)

From Figure 6.10 it can be seen that there is a requirement for some terrestrial radio links (e.g. A, B and even C) and also some satellite links (I, J and K). Figure 6.11 shows a hypothetical block diagram for such a future system. The frequency band is still to be defined (probably at the next two World Radio Conferences).

7 Terrestrial Backhaul and the Aeronautical Telecommunications Network

Summary

This chapter covers backhaul interconnection between mobile base stations and the core parts of a mobile network ground infrastructure. Broadly speaking, terrestrial backhaul has been categorized as using cable backhaul, private networks usually owned and operated by service providers (for example, the Aeronautical Telecommunications Network, sometimes called the Aeronautical Fixed Telecommunications Network, ATN or AFTN), PTT-offered 'services' and individual terrestrial radio links (these could be microwave line-of-sight links and UHF and VHF links in some of the more remote or developing parts of the world). Finally, for longer range, nowadays there is a tendency to use very small aperture terminals (VSATs) for developing or oceanic territories.

7.1 Introduction

One element that is often forgotten or underestimated when considering a mobile communication system as a complete entity is the distributed nature of base stations, hubs and mobile users. Sometimes this may be a very compact network, such as a system primarily catering for an airport area. For the national networks, it usually involves base stations in remote areas with little or no infrastructure immediately available to connect them back to the main part of the network. When engineering these systems, various options must be considered in terms of functional requirement (similar to the mobile dimensioning a network, capacity, range, format, quality), reliability, total cost, control/ownership of a network, management and maintenance as well as a time frame for realizing the network infrastructure (Figure 7.1).

Aeronautical Radio Communication Systems and Networks   D. Stacey
© 2008 John Wiley & Sons, Ltd

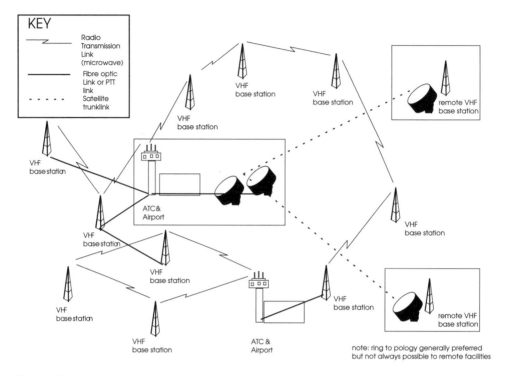

Figure 7.1 Backhaul for a mobile network.

7.2 Types of Point-to-point Bearers

Initially, it is important to understand the presentations of the interface required between a base station and the core part of a network.

7.2.1 Copper Cables

If the distance between a base station and the central equipment for a network is reasonably short (i.e. up to a couple of kilometres), copper cables can usually be used. These would support voice circuits or channel circuits for data and also the signalling and management overhead to control the base station. There could be many different formats of the copper cables.

For a few circuits these could be analogue four-wire E&M circuits, which are a six-wire system (Figure 7.2) (two for transmitted signal, two for received signal and two wires for the ear and mouth – hence E&M – signalling circuits). The E&M circuits work using DC voltages switched usually between 0 and –48 V. The two-wire E&M circuit is a four-wire-reduced version of this, where Tx and Rx paths are coupled onto the same conductors. Other set-ups also exist, consisting of analogue circuits with the signalling embedded over the channels using tone multi-tone frequency dialing (MTFD) (these can be four- or two-wire circuits). For two-wire or four-wire E&M circuits the range between base stations and the main network is usually a few kilometres. Beyond this, equalization of the cable and the signalling thresholds for E&M circuits become more involved and it is easier moving to a digital connection through a multiplex at each end.

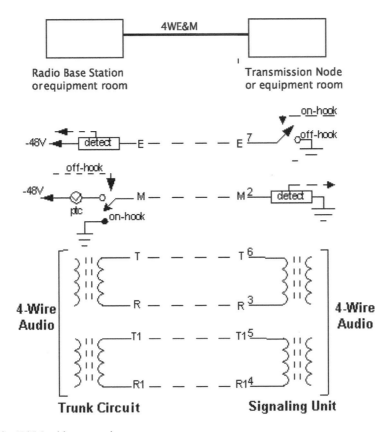

Figure 7.2 E&M cable connection.

7.2.2 Frequency Division Multiplex Stacks

In olden days multiple analogue circuits could be 'grouped' and 'supergrouped' in a frequency division multiplex stack to trunk communications between two hub nodes (Figure 7.3). It is now rare to see this 'old' set-up, particularly in the Western countries, as most telecommunication systems are moving to the technologically superior time division multiplexing (TDM) or digital style systems, which are easier to set up because no equalization has to be performed manually; therefore, maintenance and management efforts can be reduced.

7.2.3 Newer Digital Connections and the Pulse Code Modulation

The newer Pulse Code Modulation digital systems which are generally deployed in trunk networks, the E1 standard (2048 kb/s line) with 32 TDM 'slots' has emerged as an international telecommunication *de facto* standard interconnecting trunk sites (Figure 7.4). (In the United States, they have a 1.5 Mbps equivalent called a T1, but this is not so common, and in principle, the workings are almost identical and the payload has slightly fewer slots.) The E1 (T1) tributaries form the basic building block for the plesiosynchronous digital hierachy (PDH) as it is referred to. This involves an interestingly flexible way of synchronizing the network from a clocking hierarchy (depending on what equipment is available). PDH is particularly useful if

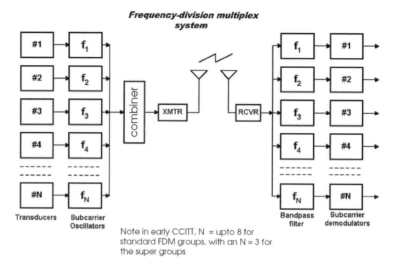

Figure 7.3   Frequency division multiplex system.

multiple base stations are located at one site. Where there are only a couple of circuits required at a site, the sub-E1(T1) rate can be used, which can be presented as 64-kbps pipes in G703 format (this format consists of two coaxial cables or can be a three-pin plug from the two coaxial cables, with the sheaths of the two cables being bonded (as earth, one for Tx and one for Rx)) (Figure 7.5).

E1 transmission is generally so cheap that it is almost used as the standard connection between outstations and also around the core of a network, even for a few trunk circuits.

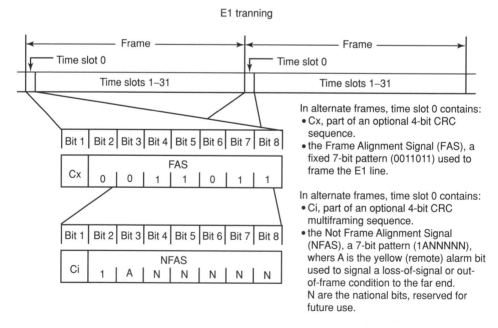

Figure 7.4   E1 format.

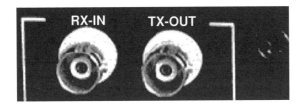

Figure 7.5 G703 format.

E1 transmission can be easily carried out across copper coaxial cables for a few kilometres without repeating. Beyond this, repeaters can be used to increase the range of the E1 links over copper cables by tens of kilometres. Alternatively, the same transmission can be carried out between multiplexers over fibre or over radio. For core parts of the network, E1 links can be multiplexed plesiosynchronously onto 8-, 34- and 140-Mbps (and the newer generation of synchronous digital hierarchy at 155-Mbps) links. Transmission network rings are often used for resilience of a network (Figure 7.6).

7.2.4 Synchronous Digital Hierarchy, Asynchronous Transfer Mode and Internet Protocol

Even PDH is now starting to become an old technology. It is being superseded by the next generation of digital trunking equipment, such as asynchronous transfer mode (ATM), synchronous digital hierarchy (SDH), Super-high-speed dense wave length division multiplexing (DWDM) working at 30×2 Gbps, and finally by Internet protocol (IP) networks, which transport traditional transmission networks in packets. Each of these are essentially able to 'transport' the existing PDH hierarchy over their payloads (backward compatibility). The newer generations of service, such as IPv6, are able to handle lots of new formats and functionalities. The trunk mobile radio channels or HF long-line site extensions can be carried within these. All these networks can be carried over copper, fibre or radio links. There are numerous other digital formats for connecting the 64-kbps line, or even sub-64 kbps rates, through an intermediate multiplex.

7.2.5 Fibre Optic

For large bandwidth connectivity, fibre optic cables are more attractive than copper cables. These can handle thousands of E1 circuits and have the added advantage of electrical isolation and immunity from lightning or ground currents. Usually E1 and all the bit rates above this can be transported by fibre.

The range of fibre systems can be up to hundreds of kilometres without repeating. This range depends on the fibre type, usually the dispersion features of the fibre, and can be as often as 10 km but more likely it is fifties of kilometres. However, for larger capacity systems, it is usual to equalize and amplify fibre circuits more often.

There are also some hierarchies used above the PDH and SDH transport layers for carrying tertiary Terra (10^{12}) Tbps line speeds. These are called wavelength division multiplexing (WDM; it is a combination of FDM and TDM) and dense WDM (DWDM). The detailed discussion of these is outside the scope of this book, but there are usually super-high bandwidth (millions of conventional voice channels) carrier-type applications using these hierarchies at the moment.

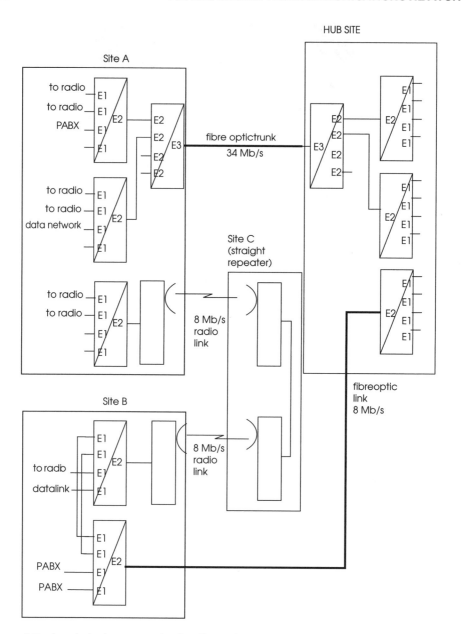

Figure 7.6 A typical private network using fibre and multiplex and trunk digital ratio.

7.2.6 Private Networks and the Aeronautical
Telecommunications Networks

This brings the topic neatly to the discussion of private networks. These are large transmission networks interconnecting a number of nodes for a large company. Most large multinational companies have substantial private networks or wide area networks. Air navigation service providers (ANSPs) are no exceptions. Such a private network would not just transport

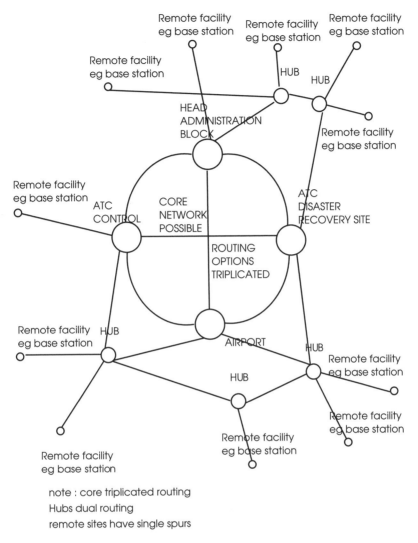

Figure 7.7 Hypothetical ATN (private network run by ANSPs).

aeronautical mobile communications but also interconnect other communication hubs for the business: These may include administrative offices and office PABXs, control centres, ATC towers and radio hub equipment rooms or remote base stations. Sometimes the ANSP's version of a private network is called an ATN or ATFN (where F stands for fixed). Normally a private network is optimized for economic cost and also for a degree of control over reliability and maintenance. Sometimes it incorporates totally privately owned infrastructure, and sometimes it leases parts of other providers' infrastructure as main or backup links (Figure 7.7).

One of the most important features of a private network is the controllability; i.e. the ANSP has total control over all elements of its network. If parts of it have been 'outsourced' to PTTs and carriers, then this control is lost, which could have operational, reliability, network management or cost implications.

7.2.7 PTT-Offered Services

In the limit, the PTT network is just a private network owned by an independent third party, some times with a billing engine/mechanism for charging for connectivity by the minute or by the month. This may include circuits presented as E1 or sub-E1, or E2, E3, etc.

The advantage of using a PTT may be that since the infrastructure is already in place between two nodes, it may provide resilience to a company's own private network (e.g. by closing a ring topology). It may provide international connectivity between two points where a private network would be prohibitively expensive.

On the downside, sometimes a PTT will not declare the routing of the service between two points, meaning its reliability cannot be easily verified and its integrity not so easily guaranteed. Experience in this field has sometimes found that even 'dual routed' circuits provided at a premium from a PTT as theoretically being independent have been found in reality to go through the same exchange or have a common power source. Usually the tariffs for providing a service can be very costly. PTTs tend to work on an internal rate of return for a period between 1 and 3 years.

7.2.8 Radio Links

Point-to-point radio links can offer an alternative to a cable point-to-point bearer. In fact, in some instances they can prove easier to install (e.g. where the terrain is mountainous), they can give better protection from third-party damage (such as cables being inadvertently dug up)/or, they can usually be installed relatively quickly (no wayleaves or trenching required), and they are usually the ideal trunking medium for the remote sites.

In the fast developing part of the world it is usual to have microwave radio bearers. These generally work in 4-, 6-, 7-, 8-, 12-, 14-, 15-, 18-, 23-, 38-GHz bands and provide hop lengths from hundreds of metres (usually at the higher frequencies) to over hundreds of kilometres (at the lower frequencies).

Before microwave frequency links existed, VHF and sometimes UHF fixed links were used to provide long hop lengths (again limited by horizon constraints but typically upto 200 km). These can still be found in some remote or developing parts of the world today. However, they tend to have poor capacity, (e.g. a couple of analogue circuits FDM multiplexed onto carrier or at best a slow speed digital link circa 32, 16 or 9.6 kb/s). For a remote repeater station this can sometimes be sufficient.

7.2.8.1 Fixed Radio Link Design

To engineer a radio link, it is necessary to perform a link budget analysis. The antenna and transceiver design parameters can usually be obtained easily from an antenna magazine (e.g. *Andrews Antennas* or *Gabriel Electronics*) and the radio manufacturer's data sheet.

A fade margin is required to make sure that the radio receive power is adequately above the threshold of the receiver. The size of fade margins typically varies from a few decibels (giving a very marginal link) to 30 or even 40 dB, which is required for a long-distance robust link. The exact fade margin required is a function of distance, frequency band (i.e. propagation mechanisms), availability required, terrain, climatical region of the world and even the relative heights of the two end sites. (This is to assess the ducting ability, which is an unwanted phenomenon when two sites have a similar height.)

The detailed designing of links can be done easily using good software tools and is not further discussed here, except for the two examples offered below. As a general rule, above 5 or even 10 GHz, rain attenuation is considered the limiting factor of a microwave link. Below 5 GHz, usually rain is not a consideration and the unavailability mechanism is usually dominated by deep fades caused by multipath refraction and reflection. An example link budget for each kind of system is given below.

Example Topology

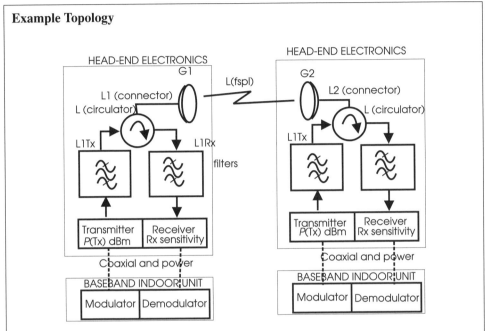

Figure 7.8 Microwave radio link design at 38 GHz.

Table 7.1 Microwave link budget at 38 GHz.

Tx power	A	23 dBm
Filter loss.	B	0.5 dB
Circulator loss	C	1 dB
Feeder 1 loss	D	0.5 dB
Antenna 1 gain	E	38 dBi
Fspl	F (for 5 km path)	137.9 dB
Antenna 2 gain	G	38 dBi
Feeder 2 loss	H	0.5 dB
Circulator loss	I	1 dB
Filter loss	J	0.5 dB
Receive power	K=A-B-C-D+E-F+G-H-I-J	−42.9 dBm
Receiver sensitivity	L (take from equipment spec sheet)	−78
dBmFade margin	M= K-L	35.1 dB

This is a reasonable fade margin for a 5 km hop in most geographic zones (bar extreme tropics), the limiting factor will be rain fades and can be calculated from CCIR/ ITU R

Example Topology

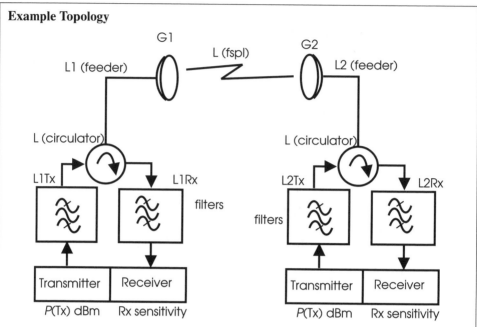

Figure 7.9   Microwave radio link design at 7 GHz.

Table 7.2   Microwave link budget at 7 GHz.

Tx power	A	29 dBm
Filter loss.	B	0.5 dB
Circulator loss	C	1 dB
Feeder 1 loss	D (assuming 30m of EW78 waveguide)	2 dB
Antenna 1 gain	E	36 dBi
Fspl	F (for 35 km path)	140.18 dB
Antenna 2 gain	G	36 dBi
Feeder 2 loss	H (assuming 45m of EW78 waveguide)	3 dB
Circulator loss	I	1 dB
Filter loss	J	0.5 dB
Receive power	K=A-B-C-D+E-F+G-H-I-J	−47.18 dBm
Receiver sensitivity	L (take from equipment spec sheet)	−82 dBm
dBmFade margin	M = K-L	34.82 dB

This is a reasonable fade margin for a 35 km hop in most geographic zones, the limiting factor will be multipath fades and reflections on such a hop and can be calculated from CCIR/ ITU R

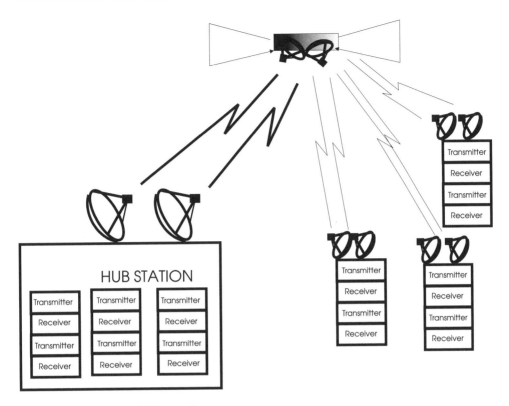

Figure 7.10   Example VSAT network.

Similarly radio rings can be established to give resilience and reliability, with critical pay-loads being dual routed at the multiplex level around a ring.

7.2.9   VSAT Networks

In the limit, where the terrestrial distances are sizeable (say over a few hundred kilometres) and where radio sites cannot readily be achieved and other terrestrial infrastructure is missing, VSATs offer an alternative to radio point-to-point infrastructure. These can be very attractive in developing countries where interconnection between multiple countries may be required (such as adjacent control centres) or in countries with multiple low-lying islands where radio links between islands cannot easily be set up (Figure 7.10).

7.2.9.1   VSAT Radio Link Budget

Again, the detail of this is outside the scope of this book, but a typical example for aeronautical network VSAT system is offered below.

Example VSAT Link

Hub part and spur part

Table 7.3 VSAT link budget.

Transmit from Hub (Site A) to Site B

Site A (Nairobi)		Modem 2 × 64 Kb/s Using Intelsat is 802 at 33°E		Site B(Dar es Salaam)	
Tx freq	14.25 GHz	Modem	2 × 64 kb/s	Rx freq	13.55 GHz
Tx power (5W)	38 dBm	Eb/No	6		
Antenna Gain (2.4m)	49 dBi	Antenna gain	17 dBi	C/No	62.98
EIRP	88 dBm	EIRP	52.7	Footprint gain	48.7 dBW
G/T	9.5	G/T	10.8	G/T	9.5
Lat	1			Lat	−4
Long	42	Long	33	Long	39
fspl	206.58 dB	C/No Satellite	115	fspl	206.14 dB
Losses	2 dB	Transponder Gain	142	Losses	2 dB
Margin	2 dB			Margin	2 dB
		Link Margin	**5.91 dB**		
Site B (Dar es Salaam)		Modem 1 × 64 Kb/s		Site B(Nairobi)	
Tx freq	14.50 GHz	Modem	2 × 64 kb/s	Rx freq	13.70 GHz
Tx power (5W)	38 dBm	Eb/No	6		
Antenna Gain (2.4m)	49 dBi	Antenna Gain	17 dBi	C/No	62.68
EIRP	88 dBm	EIRP	52.7 dBm	Footprint gain	52.7 dBW
G/T	9.5	G/T	10.8	G/T	9.5
Lat	−4			Lat	1
Long	39	Long	33	Long	42
fspl	206.72 dB	C/No Satellite	116	fspl	206.3 dB
Losses	2 dB	Transponder Gain	142 dB	Losses	2 dB
Margin	2 dB			Margin	2 dB
		Link Margin	**8.61 dB**		

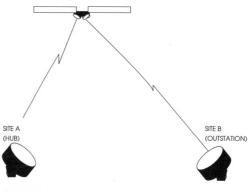

SITE A
(HUB)

SITE B
(OUTSTATION)

Figure 7.11 Example VSAT link.

7.2.10   Hybrid Network

It usually takes some or all the elements discussed above to provide a mature ATN. A hypothetical network is a combination of all the figures in this section, comprising leased PSTN lines, microwave radio links, multiplex, fiber optic and VSATs.

8 Future Aeronautical Mobile Communication Systems

Summary

This chapter looks at the next generation of mobile communications. This includes the near-term certainties such as universal access transceiver (UAT), Mode S extended squitter and surface movement communication LAN (802.xx derivatives). It also conjectures at the possibilities of wideband, code division multiple access (CDMA), time division multiple access (TDMA) terrestrial system components and various other options for a future communication system(s). It concentrates on the terrestrial parts of the system. The satellite components are mentioned in passing but are considered less mature and further away in evolution.

Near-term certainties:

- UAT;
- Mode S extended squitter;
- 802.xx LAN derivatives.

Longer term options:

- VDL extensions;
- B-VDF;
- CDMA based;
- TDMA based;
- ETDMA;

Aeronautical Radio Communication Systems and Networks   D. Stacey
© 2008 John Wiley & Sons, Ltd

- P34;
- Inmarsat Swiftbroadband.

Bands:

- VHF;
- L band;
- C band;
- Others.

8.1 Introduction

It is always very hard anticipating which direction future mobile communications will go for the aviation industry. However, there are a number of serious futuristic developments and proprietary systems on the discussion table and from here it is clear that a number of options will probably be pursued with some likely outcomes. The buzzwords at the moment have to be higher bandwidth, spread spectrum, greater mobility and flexibility, spectral efficiency, flexible air interface, application-independent air interface and band, and cognitive or tuneable software radio. Pushing all of these themes forward, will no doubt arrive at version 1 of the futuristic aeronautical communication system or systems as it is highly likely that a hybrid of a number of solutions will be taken forward, with the application layer riding independently or transparently over the top of the air interface. Hopefully this does not sound too futuristic, as from a technology perspective the ability is already there.

8.2 Near-term Certainties

8.2.1 Universal Access Transceiver

This is an initiative spearheaded by the United States. Much of the pioneering work was carried out by Mitre Corporation, starting around 1997 as a concept of a low-budget, high-bandwidth data rate communications provision.

UAT is for a point-to-point, or point-to-multipoint (broadcast mode), relatively low-range communication system working at 978 MHz. Initially it was conceived to perform mainly surveillance functions, such as ADS-B (automatic dependent surveillance broadcast) where an aircraft broadcasts its current position (or state vector). It can also operate a number of data services, for example, flight information services and traffic information services.

It can use both a structured time slot approach and a random access protocol. Each ground station covers a service volume where data exchanges can be made an aircraft configuration is shown in Figure 8.1

8.2.1.1 Frame Structure

The structure of the UAT frame protocol (Figure 8.2) uses TDMA in the TDD (Time Division Duplex) mode with regular access or random access protocols. There are called message start opportunities the bandwidth is a 1 MHz slot centred on 978 MHz, and the maximum data rate (gross) for all users is 1004.167 kbps. Inside this bandwidth each frame contains a useful

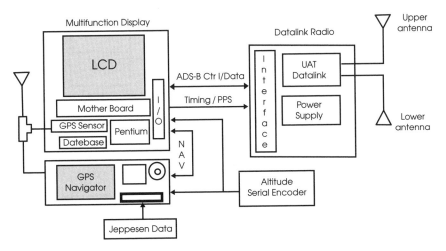

Figure 8.1 UAT system (Aircraft configuration). Reproduced from Mitre Corporation.

amount of data (or *message service block*) and there is an *overhead* to protect this or guarantee quality of service (namely FEC, code parity).

8.2.1.2 UAT Transceiver Specification

The transmitter power output can be low-powered at 10 W or for some applications high-powered at 100 W. Obviously the lower power version is preferred where multiple units are used or frequency reuse is needed in a high-density usage area, and for other areas where range is the chief consideration – for example in remoter areas – the high-power option may be preferable.

Receiver sensitivity is specified as −98 dBm (for certain conditions and probability of de-modulation). Additionally receivers can be defined as standard or high performance, where the latter is designed with enhanced selectivity to reject adjacent DME (distance-measuring equipment).

PAYLOAD AIR INTERFACE

6 ms	176 ms	12 ms	800 ms	6 ms

GROUND SEGMENT

AIR CRAFT SEGMENT
D/L
ADS-B

MSO U/L
0

MSO
3951

MESSAGE START OPPORTUNITIES (MSO) EVERY 125 ms

Figure 8.2 UAT payload air interface.

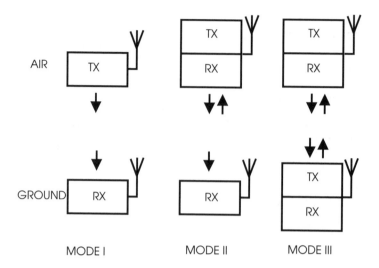

Figure 8.3 Modes of operation.

8.2.1.3 UAT Modes of Operation

There are a number of modes of operation (Figure 8.3), namely

- Mode I – airborne Tx only, ground Rx;
- Mode II – airborne Tx and Rx, ground Rx;
- Mode III – airborne and ground Tx and Rx.

8.2.1.4 Message Types

There are two message types transmitted (Figure 8.4), namely

- *The basic message block.* This is sometimes called a type 0 and is 18 bytes long.
- *The long message block.* This is 34 bytes long.

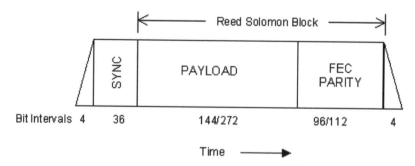

Figure 8.4 Typical message block structures.

8.2.1.5   Application and Limitation of UAT

In conclusion UAT is very much what it set out to be: a high-bandwidth, low-cost data communication provision. Its range is also to some extent limited depending on a number of factors. It
plays well into expanding the portfolio of data applications or taking away the previous barriers
of throughput limitation. Also it is semiautonomous in operation; (i.e. no central control point
exists) this has many advantages for overall system reliability.

   Limitations of UAT may include the inability to carry voice over the system, which is
currently not provided for but theoretically it could. Another potential inhibition is the fact
that at the moment only the North American region of the world and ICAO have signed up
for its usage. This could lead to the age-old problem of multiproliferation of multiple systems
and regional solutions, which cannot be a good thing for equipage and global standards. Other
than this, time will see what the future marketplace really is for UAT.

8.2.1.6   Further Reading

UAT is currently going through the ICAO standardization process and is nearing completion.

- There are two manuals on the UAT Detailed Technical Specification and implementation
  specification respectively in preparation by ICAO ACP-WG-C UAT subgroup; these can
  be viewed on the ICAO ACP website.[1]
- RTCA DO-282A.

8.2.2   Mode S Extended Squitter

8.2.2.1   Mode S Introduction

A similar high-rate communications data channel can also be embedded in secondary surveillance radar in the form of the Mode S transponder system architecture.

   Basically the secondary surveillance radar works with the ground transmitter sending out
carrier transmissions at 1030 MHz, pulse sequences (or interrogations) which are received by
the transponder (active Rx/Tx electronics) on the aircraft within the radar's service volume.
The airborne transmitters respond to the interrogation with similar known pulse sequences
based on a carrier transmission at 1090 MHz.

   The primary function of the radar is for tracking of aircraft positions and for labelling each
aircraft (this is embedded in the transponder message). From a security perspective it also
enables the air traffic control service providers and the military to identify friend and foe
(Figure 8.5).

   As with many legacy systems, it was identified early on that there was significant redundancy in the SSR (Secondary Surveillance rader) radar protocol and the ability to put additional
pulses into the existing frame structure which could be used to carry data. This is the basis of
Mode S transmissions. Of course compatibility between these new types of radar and transponders and the classical SSR radar (operating in what is classically called mode A or mode C)
is kept.

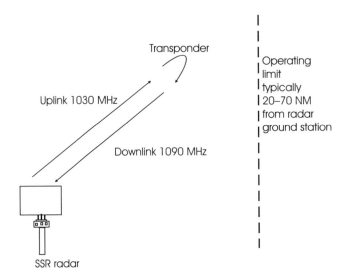

Figure 8.5   SSR radar operation and operating volume.

8.2.2.2   Pulse Interrogations and Replies

In the SARPS, a number of different interrogation pulses and replies are possible by SSR ground radar. Concentrating just on the newer mode S variant. There are long and short pulse variants for both the interrogator and the replies (see Figure 8.6)

For the short pulse, a payload of up to 56 bits can be carried by every short DPSK data train of 16.25 µs duration, which equates to a gross bit rate of 3.44 Mbps, although the net data rate is substantially less than this, allowing for the traditional SSR frame structure which extends the frame out to 19.75 ms. This equates to a net bit rate of 3.11 kbps, which again is a theoretical maximum as it contains an overhead.

For the long pulse, a payload of 112 bits is achievable over the 30.25 µs duration, giving a maximum data rate of 3.7 Mbps, although again the net data throughput is up to 3.32 kbps and includes data overhead (Figure 8.7).

Example uplink formats	Example downlink formats
UF0 Short aircraft – Aircraft burst ACAS	DF0 Aircraft – Aircraft reply ACAS
UF 4, 5 surveillance roll call	DF4, 5 roll call reply (ident, altitude)
UF11 Mode S all call	DF11 Mode S all call reply
UF16 Long burst ACAS	DF16 Long burst ACAS
UF19 Military use	DF17 1090 extended squitter
UF20 Ident request	DF19 Military extended squitter
UF21 Altitude request	DF20 Ident response
UF22 Military use	DF21 Altitude response
UF24 Extended comm. message	DF22 Military use
	DF24 Extended comma

Figure 8.6   Mode S interrogation and response formats.

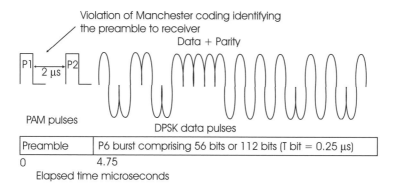

Figure 8.7 Mode S uplink data format.

An alternative method of sending communication packets is by using the extended length message protocol. This has a lower overhead and can push 80 bits through one standard frame, which can be cascaded together in subsequent packets (Figure 8.8).

8.2.2.3 *Further Reading*

For a more detailed explanation of Mode S, see

- ICAO SARPS, Annex 10, Volume IV, Chapters 3 and 4;
- RTCA DO 218B MOPS for Mode S airborne datalink processor, or;
- EUROCAE MPS 718.

8.2.3 *802.xx Family*

In the late 1990s and early 2000s it was identified that one of the areas of the air transport industry which offered potential improvement in safety statistics was that of ground incursions, i.e. the careful management of ground movements at an airport and the associated operation

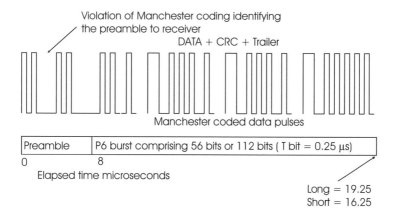

Figure 8.8 Mode S downlink data formate.

of runway(s). In contrast to the requirement to maintain levels of safety, it is also noted that a substantial growth of movements has happened and will continue to happen at some of the hub airports of the world. This makes the management of clearances, run outs and take-off and landing time slots more critical and a crunch point ultimately for airport capacity with a knock on impact into en route delays.

Regarding the management of communications for these functions, today routine movements are conducted by data exchanges in, for example, take-off clearances, but the separation and control of aircraft positions on the ground is today very much done manually by controllers in coordination with the computer tools and surveillance equipment over conventional VHF or UHF voice channels. Under adverse weather conditions such as fog, airports throughput hourly movements can drastically dive.

What has become apparent is that if more bandwidth could be made between aircraft and the ground whilst taxiing and for airport landing and take-off operations, more of this functionality could be computerized. In addition, with the communications data stream, by receiving it in multiple sensors around the aircraft a more detailed or accurate position of the aircraft could be made. This secondary function is sometimes referred to as the 'Airport Network and Locating Equipment' aspect of it, basically a surveillance function potentially complementing limited airport surface detection radar (if it exists.)

In assessing what is the most appropriate protocol for this communication channel, the usual design issues were considered.

Throughput capacity was important in what is the maximum data rate that can be reasonably achieved.

Coverage would be equally important and a high degree of availability would be needed for the airport area; this may be typically a couple of kilometres which could be covered potentially by more than one ground-based transmitter and many more ground-based receivers (and in some of the biggest airports that are outside this model, more transmitters could be deployed).

Mobility (in terms of aircraft speed) will be relatively small, with most movements being under 50 km/h. Speeds of some of the latter stages of take-off and the touchdown phases maybe approaching 200 km/h.

Considering all the above and a need to ideally adopt 'off the shelf' type technology, the 802.xx wireless LAN protocols are considered to be most appropriate and this technology is being pursued by the United States. These are expanded versions on the common radio LAN, usually known to the corporate or even domestic consumer as 802.11b, running at 2.5 GHz with a range of typically 50 m or less and a gross data rate of 11 Mbps. Obviously this particular specification is not appropriate for airports (Figure 8.9).

8.2.3.1 802.16

In particular it would seem the 802.16 protocol would be most appropriate to the airport scenario. The LAN or local area network having a range of typically ~50 m can be expanded; this gives the concept of a MAN or metropolitan area network, where the range is more like 5 km.

Originally 802.16 operates at the higher frequencies of 10–66 GHz and needs line of sight (LOS). However, a lower frequency variant operating at 2–11 GHz has been ratified as 802.16a. There is also a non-LOS standard under development, 802.16f, and two longer

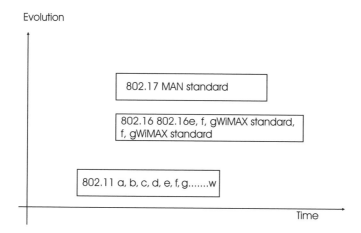

Figure 8.9 802.xx derivatives.

range versions, 802.16d and 802.16e. For the aviation scenario, work at WRC 2007 is making way for an AM(R)S provision for an aeronautical MAN to operate at 5091–5150 MHz (Figure 8.10).

8.2.3.2 Specification

The specifications for aeronautical 802.16 are still very much under development, particularly the latter versions, but the above table it should be possible to pick which aspects are most appropriate to the airport scenario.

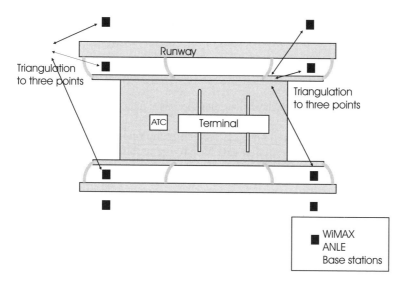

Figure 8.10 Airport MAN.

8.2.3.3   *Application and Limitations*

The opportunity for airport MANs is quite clear and these can be used to carry data and voice and many of the new higher speed data applications. The protocol is still under development and the optimal solution for aviation to adopt will still take some time to develop and standardize, and assuming an allocation at WRC 2007, some of the first airport MANs could be seen to be in operation by the end of the decade.

The ability to measure the delay from the aircraft to each of the triangulation points on an airport is further behind in the development process, which in turn will take some time in developing the surveillance functionality of the link. In fact arguably, the aircraft could just relay its differential GPS (global positioning system) position inside the MAN protocol, which could be easier to achieve at the cost of reduction in available bandwidth for other functions. This is still to be decided.

8.3   Longer Term Options

8.3.1   *Analysis*

For analysis, it is important to start with a fresh sheet of paper and literally recap and scribble all the available systems today down on the sheet. After doing this, one can add all the short-term solutions described so far in this chapter. The result would be something like as shown in Figure 8.11.

Existing systems available today	Technologies possible future systems

• VHF DSB-AM (50, 25 and 8.33 kHz)
• ACARS over VHF, HF, satellite
• VDL
• HF
• Satellite
• JTIDS/MIDS
• Telemetry links
 	
Imminent
   • UAT datalink
   • Mode S datalink
	
	• BVHF
	• CDMA
	• TDMA
	• OFDM
	• Freq hopping
	• Satellite
	• 3G/4G/5G
	• Pulsed

Figure 8.11   Existing mobile communications plus short-term futures.

Question: now comes the hard part, i.e. predicting the future. Knowing everything that is known today, what will be the requirements for a future mobile communication system?

8.3.2   *Answer*

There is no certainty as to which way this will go; however, on analysis, here are some points. Requirements for a new System.

• It is likely that the 8.33 kHz programmes, the VDL2, 3 and 4 programmes [and the UAT/Mode S extended squitter programmes] will buy some breathing space into the 2010–2015 time frame but after this the capacity will become saturated, and the technology obsolete. After this, the spectrum congestion 'brick wall' will undoubtedly return again as it did in the 1950s, the 1960s, the 1970s and 1995. This time channel splitting cannot be done without a significant rethink of the communication requirements and environment.

- *Enable software configurable radio systems.* This will be discussed later.
- *Spectrally more efficient.* As spectrum becomes more congested and a more sought after commodity, economically it is easier to strive for a spectrally efficient system than to pay the opportunity cost of taking up more spectrum. This is very much the way radio administrations are viewing spectrum allocations today.
- *Economics.* The longer term solution needs to be economically realizable both in terms of ground infrastructure roll out and air infrastructure.
- *Improvement.* Potentially a new system could open doors to better operating scenarios and techniques and do away with some of the current operating limitations.
- *Security.* This has not been a requirement in the past but post-9/11 robustness, authentication, jam resistance, integrity and encryption are clear considerations.
- *Safety.* More on-line diagnostics and real-time sharing of information of real-time flight data is required.
- *Maintainability/modularity.* Emphasis is on avionics and systems with in-built fault diagnosis equipment, to minimize aircraft turnaround times and to minimize down times; consequently, this places a requirement on some real-time data communications either directly or indirectly.
- *Easily certified.* From an airworthiness, regulatory and radio regulatory perspective.
- *New requirements.* For example, UAV requirement coming through high-altitude platforms, ANLE and new data applications.
- *Possibly change from a sector concept to a cell concept transparent to operators.* Whole new operational concept needs validating.
- *Resilience/reliability* needs to be the same if not better than existing systems. This overlaps somewhat with the maintenance and modularity concepts.
- *Sponsored.* Here is a new concept that will be described more in Chapter 9. One enabler is a financial or economic incentive to do something. If benefits can be shown to have a tangible value greater than doing nothing, a sponsor can usually be found.
- *Spectrum requirement.* This is an enabler for a new system. Provision for new AM(R)S system components are being sought at WRC 2007.
- *New operating concept.* Identify the mobile communication operating concept for the next decades. Some studies are underway within FAA, ICAO and Eurocontrol.
- *Available technologies.* Investigate new technologies available for adaptation or deployment in the current VHF band or beyond this. Emphasis is placed on plain of-the-shelf technology components needing little or minor adaptation to realize a mobile communication system, which minimizes risk and maximizes economic case for the project plan (FAA, Eurocontrol, ICAO WG C for Dec 04).
- *Other considerations.* For example, aerial farm *co-site* issues on-board aircraft and minimal equipage lists (Figure 8.12).

8.3.3 The Definition Conundrum

When all this is considered, broadly speaking, the aspects to the puzzle can be categorized into three domains. These are interlinked as the Venn diagram suggests; however, it is necessary

> + **Capacity voice**. Must be able to provide for a threefold+ increase in voice channels/decade as has been seen in past with existing latency and quality.
> + **Capacity and bandwidth data.** Must be able to provide data channels for modern applications. The data channel should not impede the application. It should cater for all existing datalink requirements and 'expanded high-speed datalink'.
> + Cater for all existing datalink requirements and expanded datalink.
> + **Mobility.** Must be able to provide for all levels of mobility, fastest aircraft speeds (Will this incorporate future rockets?)
> + **Backward compatible.** This is always an ongoing evolution of mobile communications.

Figure 8.12   Influences on a new AM(R)S system.
Note this list is a comprehensive analysis of the situation today but is likely to change or have emphasis or de-emphasis on aspects as the technology and scenerio are refined.

to decouple them to an extent and consider each of the aspects in isolation before putting the composite picture back together again (Figure 8.13).

8.3.3.1   The Requirements or the Operational Scenario

Here, one has to determine how to interpolate or extrapolate the operational environment as it is known today. There are statistics of air traffic movement available today and historically for the same months in the preceding years (see the Eurocontrol published Statfour figures). In the simplest case this growth can be extrapolated to project long term growth in the industry. For the long term an arbritary figure of 2030 is taken as a good datum to when the future system will be built to last until (Table 8.1).

This is maybe an oversimplified picture as the crunch points definitely become the major hub airports where radio communication capacity is saturated, and stacking aircraft at these points can add additional peak burden in these airport or terminal manoeuvring areas.

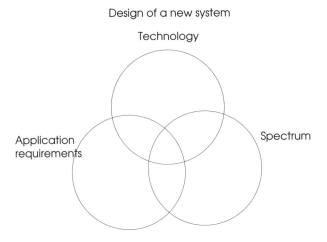

Figure 8.13   Definition conundrum.

Table 8.1 Comparison of IEEE wireless standards.

	802.16a	802.16e	802.16g
Frequency band	<11 GHz	<6 GHz	
Gross system bit rate	75 Mbps	15 Mbps	
Channelization	20-MHz channels	5-MHz channels	
LOS/non-LOS	Non-LOS	Non-LOS	
Modulation	OFDM	Enhanced OFDM	
Bandwidth	1.5 up to 20.0 MHz	Flexible, 1.5 up to 5.0 MHz	
Range	6–10 km	1.5–5.0 km	
Mobility	Fixed	Up to 60 km/h	>60 km/h, but top end undefined
Maximum spectral efficiency bps/(Hz cell)		Downlink 6, Uplink 2	
Average spectral efficiency bps/(Hz cell)		Downlink 2, Uplink 1	
Maximum throughput per user		Downlink 3 Mbps, Uplink 1 Mbps	

The next question is how will the applications change over this time? It is likely that there will be an increase in data applications whilst voice requirements arguably could reduce as a consequence or in conservatively worst case, voice plus data will be required. In what volumes? Another way to analyse this is to take a look separately at the mobile phone and data communications revolutions. This is an enigma that has been experienced in the other communication non-aviation market sectors. What is found here in an economics model is that if a communications 'channel' is made available, demand will naturally follow, and the demand if unimpeded by the channel will increase exponentially. This is what was seen in the late 1980s–1990s with the mobile communications revolution and semi-independently with the data communications revolution (WIFI, Internet, etc.), which are still going on. The argument is that even if we enable a provision for aviation communication, who knows what the limits of demand will really be!

8.3.3.2 *Technology Options and Frequency Band*

Broadly speaking, a number of different well-known technologies have been shortlisted in the studies being undertaken by various aeronautical interests. The favourites are listed out below. Also the sequence is what is considered by the author to be the most favourite pick list, most likely first – it should be pointed out that this is very sensitive to change and other political and external forces.

1. CDMA based (including wide band and narrowband variants, FDD and TDD);
2. TDMA based (this includes VDL extensions, B-VDL, new TDMA solutions or ETDMA);
3. P34;
4. Satellite solutions (e.g. Inmarsat Swiftbroadband).

Table 8.2 802.xx family.

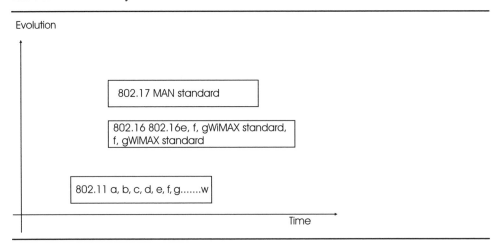

At the moment the candidate frequency bands under most serious consideration are (not in any necessary order of likelihood):

1. VHF (108–137 MHz);
2. L band (960–1215 MHz);
3. C band (5000–5250 MHz);
4. Others (maybe 2700–3100 MHz).

8.3.3.3 Spectrum Requirements

These can be broken down into a number of constituent components:

1. The propagation properties of spectrum and how suitable they are to the requirements;
2. Base bandwidth throughput;
3. Mobility;
4. Spectrum efficiency.

If the latter three are plotted on the graph with technology, an interesting observation can be seen directing us to a possible solution (Figure 8.14). This is the basis of why CDMA, TDMA, and the wireless LAN/MAN proposals are considered good candidates.

8.3.4 A Proposal for a CDMA-based Communication System

A CDMA system has been proposed as a solution to the future aeronautical communications requirement or a portion of it (Figure 8.15). This is a spectrally efficient solution that could provide for all expected future voice and data needs for the next few decades. In particular it lends itself well for flexibility, expansion options, the ability to carry multiple and high-capacity data services and voice.

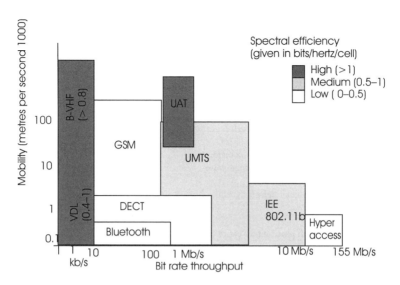

Figure 8.14   Technology versus bandwidth versus mobility versus spectral efficiency.

A number of topologies are achievable, namely FDD and TDD working (Figures 8.16 and 8.17). These can be realized in the L, S and C band regions with reasonable and economical ranges. These could in their narrowband form be achievable in the VHF band, provided some spectrum clearing could be achieved. From this sensitivity analysis (conducted at 5 GHz), it shows cell radii of upto 50–60 km to be optimal before the technology becomes spectrally

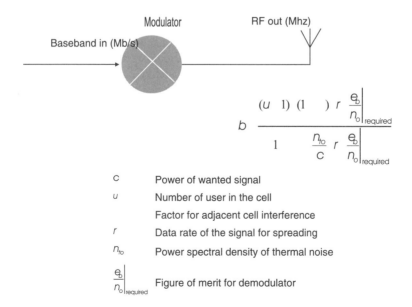

Figure 8.15   Simplified CDMA model.

From the previous equations (2.36, 2.37, 2.38 and 2.39) described in the theory section

The boundary conditions for a a CDMA system can be described by

$$B = \frac{(k-1)(1+\alpha)r\,(e_b/n_o)}{1-((N/C)r(e_b/n_o))}$$

where B = RF spectrum bandwidth
k = number of mobiles in a given cell
α = adjacent cell self interference factor
e_b/n_o = figure of merit for the demodulator
N = power spectral density of thermal noise
C = power of wanted signal
R = data rate of the signal before spreading

Typical CDMA link parameters could be

B	Either 1.75 MHz or 5 MHz per uplink or downlink
k	Typically up to 50
α	Typically 0.2–0.4 but theoretically can go as high as 6
C	Typically 20 W
R	From 16 kb/s upto 384 kb/s
e_b/n_o	Typically 4
N	kT

Figure 8.16   Link budget boundary conditions.

inefficient. This is a much smaller granularity than in the existing VHF coverage topology and suggests many base stations should give a blanket area coverage and ultimately economic cost issues. However, this is not necessarily a show stopper, it would seem that it would be beneficial rolling out such network infrastructure at congestion hotspots, for example around airports and in TMA areas and maybe in the air corridors of 'Core' Europe. This could then supplement and in some instances replace the VHF network.

Some analysis to the sensitivity of the system parameters gives the following results (Figure 8.18).

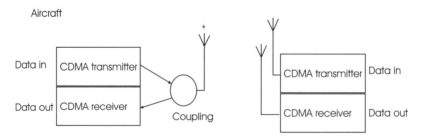

*Note: Can be the same antenna for TDD working and FDD working with careful filter design.

Figure 8.17   Functional diagram.

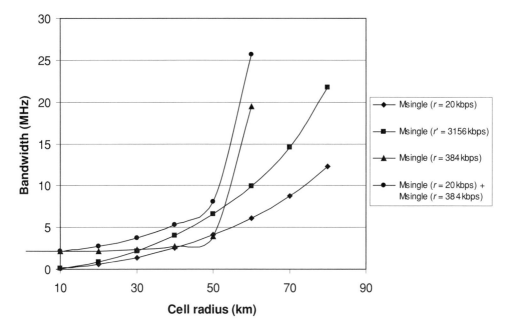

Figure 8.18   Sensitivity analysis.

8.3.5   Software Defined Radio

Today most waveforms can be generated by a digital computer chip with a finite number of discrete levels and smoothed with a filter circuit (Figure 8.19). Extending this theme, it is conceivable that VHF DSB-AM or HF SSB modulation or even CDMA TDD modulation can be performed entirely on a computer chip and passed directly to the final RF stage for amplification before being transmitted (Figure 8.20).

Conversely, the demodulation process is also possible entirely inside a computer chip. This technology could be extended to having one avionics box that could perform all different kinds

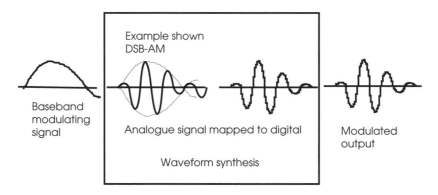

Figure 8.19   Oversimplification of computer waveform synthesis.

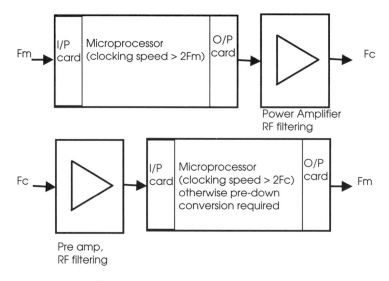

Figure 8.20 Computer modulation and demodulation.

of modulation (e.g. VDL protocol, DSB-AM protocol, 8.33 protocol, SSB protocol, future comm. Protocol) and pass it to a number of selectable RF amplifiers (Figure 8.21).

This is not as far-fetched as it might sound as the military already have this kind of equipment today. If this potential is exploited to its fullest, for the future aeronautical communications systems (and for replacement of legacy ones), potentially there would be the ultimate flexibility. A minimal 'one box' avionics (or equivalent ground-based system), minimal extra antenna requirements and the ability to send an application across an air interface based on availability of link and independent of frequency.

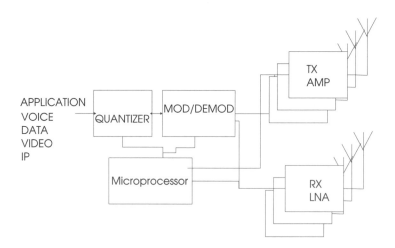

Figure 8.21 Software defined radio.

There is even talk of the next generation of software or cognitive (intelligent radio), where applications are just routed by best available air interface. This has some serious implications for the future direction of aeronautical mobile communications and even more so for the radio regulatory frame work today.

Further Reading

1. www.icao.int/anb/panels/acp

9 The Economics of Radio

Summary

This chapter provides a description of the commercial basis for realizing aeronautical mobile infrastructure. It is not meant to be a shimmering example of pinpoint precise accountancy, as it has been written by an engineer, but rather it is meant to provide the engineer or project manager with some lateral tools for justifying project expenditure and to aid economic analysis of various engineering proposals.

9.1 Introduction

So far, this book has dealt with all the theory and system building blocks for realizing mobile communication infrastructure and systems. Now at the start of the practical part, the first stage is to look at the costing and economics of realizing aeronautical mobile infrastructure.

A purist engineer would probably initially retort to this saying, 'Money doesn't come into this, you have no choice! A radio system must be realized!' (Figure 9.1).

However, it is fair to say from a commercial perspective, in order for something to happen, there needs to be a strong commercial basis for this or a commercial incentive – otherwise nothing happens. This is true for engineering at its best – the compromise between the economic situation and the best technical solution possible (Figure 9.2).

9.2 Basic Rules of Economics

Simplifying the discussion above, it can be broken down to two fundamental rules of radio economics:

1. *Every* new implementation or innovation must have an economic basis.
2. For viability, *return* should be greater than capital *investment* cost plus running cost.

Aeronautical Radio Communication Systems and Networks   D. Stacey
© 2008 John Wiley & Sons, Ltd

Figure 9.1 'You have no choice!' (cartoon of telephone lead emanating from the cockpit).

9.3 Analysis and the Break-even Point

Concentrating on this second rule, return > investment, and initially ignoring the cost of money or interest rate, consider a system that costs A million euros to install capital once-off cost. Assume the same system costs B million euros a year in total running/maintenance costs. Now consider the cost of not implementing the above system and let the initial capital cost expected as C million euros once-off cost, with a revenue cost of D million euros for not implementing a system. Now consider the number of years a system is intended to be in place and call this y. A simple equation governing economic viability is $C + (y \times D) > A + (y \times B)$. The point in time at which the left-hand side expression equals the right-hand side is called the break-even point (Figure 9.3).

9.4 The Cost of Money

The system is further complicated by the cost of money. That is, putting money 'up front' costs more in real terms than deferring the same actual expenditure to a later date.

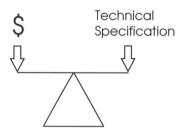

Figure 9.2 Balancing economics with best engineering.

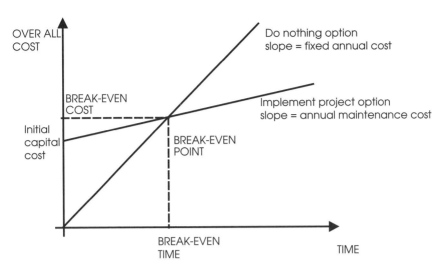

Figure 9.3 Break-even point graph.

9.4.1 Some Basic Financial Concepts

Future value is the amount that a deposit or series of deposits will grow to be over time at a particular given interest rate.

Net present value (NPV) is the total amount that a series of future payments is worth now. This is the reverse of the future value.

$$\text{NPV} = \sum_{t=0}^{T} \frac{\text{CF}t}{(1+r)^t} = \text{CF0} + \frac{\text{CF1}}{(1+r)^1} + \frac{\text{CF2}}{(1+r)^2} + \cdots + \frac{\text{CF}T}{(1+r)^T}$$

where CFt is the cash flow at time t and r is the interest rate or cost of capital.

Internal rate of return provides a measure of the average annual rate of return that a project will provide. If the internal rate of return exceeds the required return for the project, the project becomes viable and should be accepted.

Example 9.1

Consider that a new radio repeater site is being built to accommodate a VHF air/ground base station. The choice of routing the trunks from these base stations back to ATC central control is by either a fixed leased line provided by the state PTT (Post, Telegraphy and Telecommunication Organization) telecommunications authority.

For this the authority quotes a € 20 000 install cost and a € 45 000 a year lease cost, payable yearly in advance.

Alternatively, a feasibility study for a microwave radio link shows a link can be installed for € 90 000 capital cost plus a revenue cost of € 10 000 a year payable in advance each year.

Ignoring all technical considerations and an interest rate of 8 %, what are the economics of the two options. Assume either solution has a zero scrap value at any point.

Answer

Table 9.1 Simple viability analysis.

| | Cash flow | | | | |
Time (Yr)	Option 1: Leased lines	Option 2: Microwave	Actual cash flow	Discounted cash flow: Cash flow/$(1 + r)^t$	NPV
0	65	100	−35	−35	−35
1	20	10	10	9.26	−25.74
2	20	10	10	8.57	−17.17
3	20	10	10	7.94	−9.23
4	20	10	10	7.35	−1.88
5	20	10	10	6.81	4.93

Conclusions

- This demonstrates that from an economic perspective the microwave link solution is best in the long term (usually the case for aviation) and pays for itself over 5 years (i.e. when NPV turns from − to +).
- If a shorter period is required, the leased line is a better economical option.

9.4.2 Inflation

This model so far ignores inflation (i.e. how the lease and maintenance costs go up every year). It can be done by adding a $(1 + i)^t$ function in the denominator in the NPV equation as required.

Example 9.2

Consider a transatlantic aircraft fitted with the Inmarsat Aero satellite phone system.
Assume the cost of equipment and installation is € 100 000. Assume that each phone connection costs the airliner or business € 2/min.
In order for them to recover the cost of the investment within a year, what call rate must they offer to the customer? Assume each day the flight returns and expects to have 100 minutes of satellite phone calls. Assume billing is done monthly, cost of borrowing money is 8 % and inflation is 4 %. Assume income is received at the end of each month.

Answer

Let tariff charged per minute $= x$
So income per month $= (x − 2) \times 30 \times 365/12$
Monthly interest rate $= 0.08/12$
Monthly inflation rate $= 0.04/12$
Manipulating the value x in the following spreadsheet until after a year the NPV $= 0$
Table 9.2 is generated using tariff of € 5. In the 12th month, it just comes positive.

Table 9.2 Viability analysis using compound interest.

Month	Cash	DCF	NPV
0	−100 000	−100 000	−100 000
1	9125	9034.45	−90 965.55
2	9125	8944.81	−82 020.74
3	9125	8856.05	−73 164.69
4	9125	8768.17	−64 396.51
5	9125	8681.17	−55 715.34
6	9125	8595.03	−47 120.32
7	9125	8509.74	−38 610.57
8	9125	8425.30	−30 185.27
9	9125	8341.70	−21 843.57
10	9125	8258.93	−13 584.65
11	9125	8176.97	−5407.67
12	9125	8095.84	2688.16

DCF, discounted cash flow.

9.5 The Safety Case

Sometimes engineers or operational personel quote the safety case as a basis for installing a piece of equipment or infrastructure regardless of the economic basis. This is not a sound argument! Every project needs some kind of economic basis. With a safety system, sometimes the economic justification is less obvious or less tangible.

Some lateral thoughts on this suggest that if it is a new safety system, the cost to install and run it needs to be less than the existing costs for existing infrastructure or the cost for having no infrastructure. Another solution may be to obtain the cost from bigger project (e.g. mandate it as part of an aircraft fit out such as 8.33 kHz working, get the European Union to understand this cost, maybe get them to underwrite the economic savings to Europe or absorb the costs in the aircraft purchase/manufacture cost).

This brings up the idea of a 'sponsor'. This is another way of implementing a safety system: to find an interested party with some money, where an economic return is provided by a related aspect.

Some examples of this could include the following:

- The Connexion by Boeing system for airline passenger communications (theoretically sponsors sales of Boeing aircraft or products).
- Satellite phones on-board, providing the multifunctions of airline passenger communications (APC), air traffic control (ATC) and airline operational communications (AOC) communications to the aircrew.
- VDL radio systems and 8.33 kHz.
- The next generation of communication system.
- Weight savings – the estimated cost of 3–5 L of fuel to carry 100 kg every 100 km.
- Delays cost money to airline operations. The cost estimated by IATA was 5700 million euros in 1999 (for Europe) (estimated 75 % borne by passenger/business and 25 % borne

by operator). These are 150 % of en route charges. Can the non-implementation of a system have delay consequences that can be measured?

- Landing slots at some of the major airport hubs have a financial price tag put on them, which is beneficial to the airport operator. Can the non-deployment of a solution jeopardize this?
- Not meaning to sound morbid, but ICAO and IATA have statistical values for probability of accident of an aircraft based on global averages, sector and region. Airlines also have a financial value of what insurance payout is expected for passenger and crew death or injury – if a new radio feature can increase this safety statistic then simple expectation calculations (leave the detail to the actuaries) can yield an economic benefit.

9.6 Reliability Cost

Particularly pertinent to the last of the above bullets, there is a concept that reliability costs money. So a trade-off is often required, i.e. *reliability* versus *cost*. In many cases, aeronautical mobile systems are duplicated and can even be triplicated on a safety requirement basis. A good example of this is the minimum equipment list. This is a minimum list of functioning equipment an aircraft must have (mandated under State law) to be allowed by the CAA of that country to take off. If a piece of equipment is not working, it can jeopardize an airlines operation or scheduled flight (which can impact onto the next schedule and so on) for which economic losses get very big very quickly. By simply duplicating some equipment for a 'fractional' extra cost and the cost of carriage, these financial penalties can be avoided.

Example 9.3

Assume that the limiting case for an airport communications radio infrastructure is the cable between the ATC control room and the base station radio room about 2 km across the airport (i.e. the reliability of equipment is ideal at 100 % and can handle multirouting connections).

The availability of this single link is put at 95 %; however, with ATC communications, an availability of 99.9999 % is required. An option is to run additional duplicated feeder(s) by different routes.

Also assume the capital cost of the overall installation is negligible in comparison to the cross-airport link. Cross-airport links can be assumed to cost € 50,000 each.

Calculate the cost versus reliability of the system for duplication, triplication, etc.

Answer
Conclusions

Table 9.3 Reliability versus cost.

	Availability	Unavailability	Downtime per month	Cost 1000€
Single system	0.98	0.02	14.6 h per month	50
Duplicated system	0.9996	0.0004	17.5 min per month	100
Triplicated system	0.999992	8×10^{-6}	21 s per month	150
Quadruple system	0.9999998	1.6×10^{-7}	0.42 s per month	200

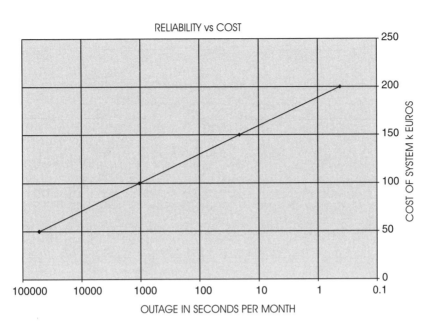

Figure 9.4   Reliability versus cost.

- Increased reliability costs money.
- There is a trade-off between cost and reliability.
- Is 20 seconds of uptime on a safety system really worth an extra € 50 000?
- Again requirement for increased reliability needs economic justification (i.e. MTCE savings, turnaround time save, no lost downtime from aircraft grounding or aircraft stacking, or improved system payout expectation.)

9.7   Macroeconomics

To put all this in perspective, the ANSPs (Air Navigation Service Providers) spend in the aviation industry is a very small component of the total industry. Of this ANSP spend, the ground infrastructure spend for communications is also a very small percentage.

In 2006, the world economy is estimated to be worth somewhere around 30 000 billion US dollars. Of this the transport sector represents about 6 % of GDP. In comparison with the latest IATA figures, the airline turnover is approximately 80 billion US dollars for Europe, of which the operation cost is somewhere between 60 billion and 80 billion US dollars for Europe.

From these models, it is possible to put financial values on high-profile routes (e.g. Frankfurt to New York JFK), to put financial values on peak time landing slots at the major airports. (These are worth millions Euros per anum for just say the 9 a.m. slot on a Monday morning at London Heathrow).

As already discussed, delays to landing or in getting to terminals cause a huge build-up in cost. These costs are often the real reason for realizing new procedures or new equipment instead of what is published – the hearsay answer of 'to improve safety'.

10 Ground Installations and Equipment

Summary

This chapter deals with the practicalities of realizing ground-based infrastructure, in particular all the component parts of a radio system plus the equipment room, building services, civil and tower plus associated services. It goes through some of the high-level considerations necessary for successful project implementation and operation.

10.1 Introduction

Generally speaking, this chapter looks at physically installing, operating and maintaining the equipment and systems described in Section B (Chapters 3 to 8).

Firstly consider the environment of where the equipment operates and what are reasonable boundary constraints within whose envelope the equipment needs to successfully operate.

10.1.1 Environment

10.1.1.1 Indoor Environment

Temperatures. Most equipment is installed in purposely built 'friendly' indoor equipment rooms with stable temperatures, ventilation and humidity. However, the worst-case scenario is when the climate control for such a room is lost for an amount of time. In the tropics or desert, particularly for remote facilities, this could mean rooms reaching temperatures of +40 or even 50 °C very quickly, and usually with the power components of the equipment at much higher temperature (in some cases it can be known for equipment such as power amplifiers or power supplies where heat dissipation is substantial to go over 100 °C).

Conversely, in the polar, continental regions it could mean that the room ambient temperature gets to −40 or −50 °C, although normally the electronics will remain slightly warmer from their own power drain. A very typical environmental temperature range specified for aeronautical electronic equipment is −20 to +55 °C; the equipment shelter should be designed to keep well

Aeronautical Radio Communication Systems and Networks   D. Stacey
© 2008 John Wiley & Sons, Ltd

within these limits (even under heating, ventilation and air-conditioning failure for prolonged periods).

Humidity. This seems to be a lesser problem, but obviously there comes a point when it is too humid and condensing starts to take place inside an equipment room or inside parts of equipment; at this point microelectronics can short-circuit. Also longer term effects such as corrosion or build up of moisture in places is also problematic and should be avoided. As a general rule, of thumb, humidity should not be allowed to go beyond 80 % at the ambient temperature.

To protect against humidity an active dehumidifier may be required. In the short-term (such as when electricity is lost and consequently dehumidifying or air conditioning), water-absorbing crystals can be left in an equipment room – these will turn colour when they are used, which indicates a humidity problem.

Protection. Any extreme environmental condition can cause temporary failure of equipment and in some cases more permanent failure; it also undoubtedly prematurely ages equipment, resulting in faster acceleration through the reliability bathtub curve, described in the reliability section (Chapter 2), earlier failure and indirectly increased cost.

It is important to be aware of the environmental specification for equipment rooms and the likelihood of having outages. Also at those times when these will be out of specification, and if necessary or feasible either perform the necessary upgrades to the equipment room or take the consequential engineering precautions with the equipment.

10.1.1.2   Outdoor Environment

Now, consider outdoor environment, where the climate is usually even more extreme.

Again *temperatures* can swing cyclically from day to night, from season to season and from day to day. Equipment must be appropriately rated for this. In some extreme desert areas, it is possible to see temperature extremes of $+50\,^{\circ}\mathrm{C}$ and down to $-50\,^{\circ}\mathrm{C}$, although usually not within 1 or 2 days.

Wind speeds, particularly on radio sites, by their very nature can be above the regional averages and full account must be taken of what the highest wind speed is likely to be in a given time. (Some countries call this the 100-year wind speed or 50-year wind speed and multiply this by a safety factor; in other places this information is not available but considered estimates can be made.)

Humidity for different regions can vary; e.g. at the poles it stays almost constantly at zero, whereas conversely in certain tropical areas it wavers almost permanently above and below 100 % and in other areas it cycles somewhere between these. This can have a very significant effect on equipment, particularly joints, seals, parts of high expansion or sensitive to corrosion or water build-up. Again it is important to know the parameters applicable to the area and to the installation.

UV exposure is a necessary consideration, especially for plastics that can become brittle with time and may need protecting by oversheathing or covering, for example.

Weight loading can be a serious issue for towers. It should be noted that the tower wind loading is usually the limiting factor of tower capacity, not its static weight-bearing ability. Extending this principle, under poor weather conditions when ice build-up can easily quadruple wind forces on a tower or overhead gantry by providing an increased surface area. For the wind force when water expands, the phenomenon of 'burst pipes' can also be applied to hollow tower legs and waveguide ducts. In the limit ice-loading forces can distort towers, ladders, gantries and damage waveguide runs and even break connections.

This weight loading can have some serious safety implications, as it is not unknown for parts of towers to fall off under such adverse conditions; as towers 'thaw' or icicles spear through unprotected parts at the basement. This happens all too often and it can be quite common in some environments to apply the hard-hat policy around towers for this reason (not that hard hats would necessarily protect from the extremes of this). The heavily ice clad tower shown in Figure 10.1a is probably in a very unlikely place. It is just behind Canberra in Australia (Figures 10.1b and 10.1c).

Lightning if it does not conduct seamlessly to ground from where it strikes on a tower or building, can cause untold structural damage, but more likely it can incinerate some of the

Figure 10.1a   Example of a heavily loaded tower (Black Mountain tower, Canberra).

Figure 10.1b Example of a remote VHF facility (Remarkables, New Zealand, this site has helicopter access only).

Figure 10.1c Central trunking site (Black Mountain, Canberra, Australia).

more sensitive and expensive components of a radio system when it gets into the equipment room such as sensitive low-noise receivers (usually used to deal with picowatts of power or less), other radio equipment and power supplies as well as melting large conductors or waveguide and cable runs.

Taking into consideration the environment limitations, a number of good practices can be used to compensate for each of these aspects or hazards and to reduce the likelihood of having problems, such as a good earth mat, regular bonding of tower parts and antennas at regular intervals, waveguides and feeders should be bonded at top and bottom of tower and at bends with minimal change in direction of earth bonds.

Ingress of *dust and sand* into joints that are then vibrated by the wind can eventually abrade and break.

To summarize, it is important to have considered all of the aspects listed above in relation to the equipment to be installed and operated. In many cases, good local codes of practice can be found as to installation guidelines (from the manufacturer of equipment, from the organization who is operating the equipment or in national codes or specifications) can be found to ensure a successful installation and operation of equipment and a good quality of workmanship that will last the test of time.

10.2 Practical Equipment VHF Communication Band (118–137 MHz)

10.2.1 VHF Transmitters

There are numerous manufacturers of this equipment for installing at a typical ground station, good examples would be Rohde and Schwarz, ITT, Rockwell Collins, Motorola, Icom (Figure 10.2).

Typical specifications would be for the following:

Frequency: 118–137 MHz;
Mode: DSB AM;
RF power output to feeder: typically 20–50 W;
Channelization: 25.00, 8.33, 50.00 or 100.00 kHz working;
Power supply: −24 or −48 V DC;
Standard 19 in. or ETSI rack mounting (e.g. case 160 × 34 × 271 mm);
Harmonic suppression: typically 40-dB below fundamental;
Frequency accuracy: typically 5 ppm;
Environmental temperature: −20 to +55 °C;
Trunk interface (e.g. 2-W, 4-W E&M);
RS 232 or 422 remote control of receiver.

10.2.2 VHF Receivers

Again, there are numerous manufacturers of this equipment for locating at a typical ground station, good examples would be Rohde and Schwarz, ITT, Rockwell Collins, Motorola, Icom (Figure 10.3).

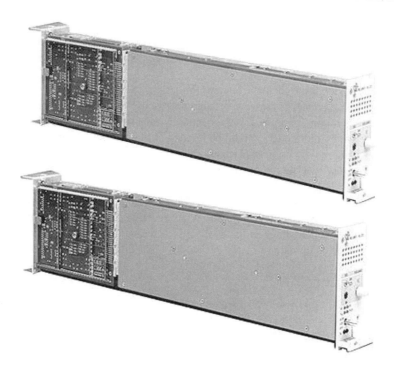

Figure 10.2   Typical rack mount transmitter. Reproduced from Rohde and Schwarz.

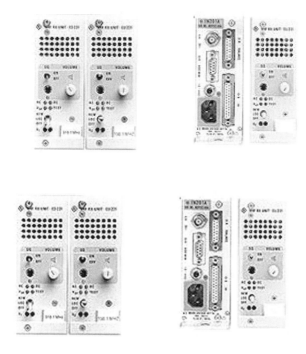

Figure 10.3   Typical rack mount receiver. Reproduced from Rohde and Schwarz.

Typical specifications would be for the following:

Frequency: 118–137 MHz;
Mode: DSB AM;
RF Rx sensitivity at Rx input port: typically −100 dBm (with 30 % depth of modulation @ 1 kHz);
Spurious response: −80 dB;
IF rejection: −60 to −80 dB;
Frequency accuracy: typically 5 ppm;
Squelch dynamic range: 3–50 μV;
MTBF 10 000 h;
MTTR 15 min (this is the once on site component);
Trunk interface (e.g. 2-W, 4-W E&M);
RS 232 or 422 remote control of receiver.

10.2.3 VHF Transmitter/Receiver Configurations

There are many different ways of configuring a base station site. Ultimately it depends on the location of a site and how many transmitter channels and receiver channels it is operating.

10.2.3.1   VHF Single-channel Dual Simplex Station Site Configuration

In its most basic form, a one-channel site operating the same transmit and receive frequency (dual simplex) is probably the most simple configuration and the transceiver can even be a mobile or portable unit (Figure 10.4). Only one antenna or one tower is needed with push-to-talk

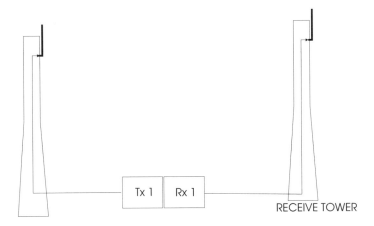

TRANSMIT TOWER
    (Note this is an example. The two antennas can be placed on one tower or even coupled onto the same antenna.)

Figure 10.4   Single-channel VHF ground station.

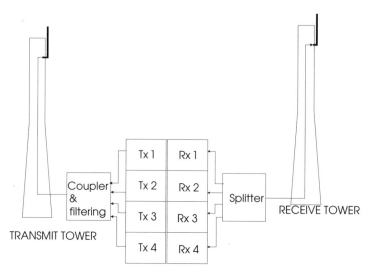

Figure 10.5 Example multichannel, duplicated VHF ground station.

switching (which can be driven remotely) to switch antenna between Tx and Rx mode, provided the site is to be operated in dual simplex mode, consequently only one tower or even a more basic mast.

10.2.3.2   VHF Multichannel, Duplicated Base Station

The considerations necessary for a typical multichannel, duplicated radio site are more complex (Figure 10.5). Firstly, it is usually necessary to separate the transmit and receive antennas (to avoid co-site swamping of the receivers by the transmitters even though they are on different adjacent frequencies; this is sometimes called the co-site issue); alternatively a receiver filter chain can be used.

Secondly, it is possible (but not compulsory) to operate one antenna for each the transmit direction and the receive direction. Transmitters can be coupled onto this, and receivers are fed from this via a low-loss splitter (minimum 3 dB per pair split); cavity filters are usually required in both paths. In the transmitter side it is to remove harmonics, while in the receiver path it is to remove spurious and adjacent channel signals into the wanted receivers or also risk of harmonic products.

Of course, one antenna could always be installed on the tower for every transmitter or receiver, but this is cumbersome in terms of multiple feeder runs and usually more expensive for fractional little increase in performance. It is sometimes done though from a reliability standpoint or where receive level is low and critical (i.e. the link budget will not tolerate a multiple 3 dB splitter loss).

10.2.4   VHF Cavity Filters

These are resonant tubes, usually with very high Qs giving a pass-band just in the 25 kHz, of concern for the channel and introducing relatively low loss (less than a dB) into the

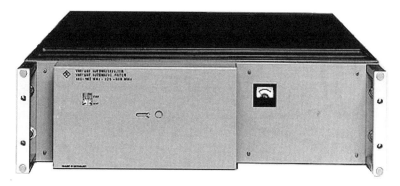

Figure 10.6   Typical cavity filter. Reproduced from Rohde and Schwarz.

feeder chain. As discussed above they are a form of pass-band filter appropriate to provide good blocking to harmonics, spurious and the other channels coupled onto a feeder system (Figure 10.6).

10.2.5   VHF Combiner, Multicouplers, Switches and Splitters

Combiners and multicouplers are often used on the input to a transmit antenna, to couple multiple transmitters. The important parameters are low loss over the pass-band (usually less than half a dB), low VSWR (voltage standing wave ratio).

Couplers can also be used to connect a main and a standby transmitter (with the input to the transmitter switched), although this is usually accomplished through an RF coaxial switch (Figure 10.7). The coaxial switch is usually controlled by a relay with DC bias (Figure 10.8). Under normal conditions the relay is energized and couples the switch to the main transmitter. Under fault or no current conditions (i.e. when told to switch from control centre), the relay defaults to standby transmitter.

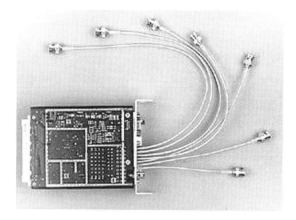

Figure 10.7   VHF coupler. Reproduced from Rohde and Schwarz.

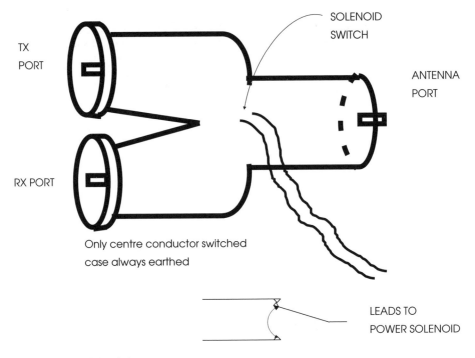

Figure 10.8   Coaxial switch.

Splitters are used to branch a receive path into two or more ways. Inevitably a small loss is incurred in the splitter (typically 0.1 or 0.2 dB), plus the consequential loss of power sharing (for even two-way split this is of course power halving or a 3-dB power loss). Sometimes splitters can be non-uniform power splitters (i.e. 1 dB is experienced one way and 10 dB is experienced the other way; this is sometimes called a 'sniffer' where the lower powers can be 'syphoned' off for measurement or monitoring purposes, leaving most of the power conducted to the receiver or transmitter where it is needed to maintain a reasonable fade margin (Figure 10.9).

10.2.6   Other Radio Equipment

10.2.6.1   HF

There is at least one HF receiving station for each of the MWARAs (described in Chapter 7). Usually there is significant separation between the transmitting and receiving functions to master the co-site aspects. (The stations are usually operated by the federal aviation administration or sometimes the radioagency on their behalf, for the data services ARINC usually have a separate facility.)

At a typical HF receiver station (there is at least one dedicated receiver for each of the channels operated by the MWARA station), sometimes the radio portion is duplicated, although it is also possible to share the protection HF receiver with the primary receivers in a 1 for N type protection topology. There are typically a couple of HF receive antenna systems feeding

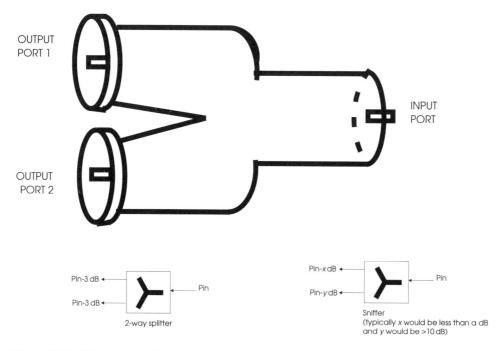

Figure 10.9   Splitters and sniffers.

the combination, which can be direct feeds or split to different receivers (Figure 10.10). Usually the receivers have band pass filters preceding the demodulator to protect from spurious, intermodulation products and out of band excessive powers that threaten the receiver under consideration.

The complementary transmitters and transmitting antennas are normally located at a separate site for the co-site issues, as already discussed.

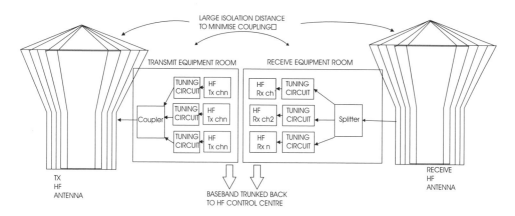

Figure 10.10   Typical HF receiver configuration.

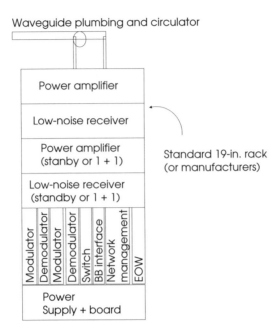

Figure 10.11   Typical indoor point-to-point radio equipment.

10.2.6.2   Microwave Point-to-point Equipment

This equipment is usually standard rack-mounting equipment (multiple rack units with height of 19 in. or ETSI rack) (Figure 10.11); it usually has standard environmental specifications and rear mounting interface points, for example, data connections and power. Waveguide terminations are usually made above the equipment and usually need some support from the overhead cable gantries.

For higher frequency equipment, sometimes only the baseband equipment is located in the indoor equipment room with IF cables (usually coaxial) going to a head-end RF mount electronics unit. For the detail, it is recommended to look at the equipment specification and installation manual provided by the manufacturer (e.g. NEC, Philips, Fujitsu, Nortel, Siemens, Nera, Nokia are manufacturers).

10.2.6.3   Satellite Equipment

10.2.6.3.1   Mobile Network

For the earth station for the Inmarsat (or similar) ground station, the equipment is very specialist, involving parabolic dishes greater than 7 m in diameter and all the tracking and control equipment associated with this, together with the electronic transmitter and receiver circuits. A basic overview block diagram is given but the detailed configuration is complex and can vary and is outside the scope of this book (Figure 10.12).

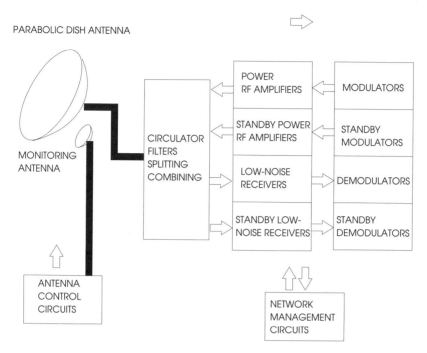

Figure 10.12   Earth station block diagram.

10.2.6.3.2   VSAT Network

The indoor equipment for a VSAT hub or outstation is similar to that for a high-frequency microwave system. The baseband unit is installed inside. This unit is typically 19 in. or ETSI rack mountable with connections to the rear. IF cables also connect to the rear. The head-end power electronics for the transmitter and the low-noise amplifier that start the receiver chain are either usually immediately on the dish as a low-noise block or immediately behind it (Figure 10.13).

10.2.6.4   Voice/Data Termination, Multiplex and Other Line-terminating Equipment

For typical rack mount radio transceivers (VHF, HF or microwave), the interface point is usually voice channels (as coaxial), 4-W E&M circuits, or RS232 or D type connectors with various physical and electrical presentations. This needs to interface with either multiplex or long-line cable extensions or increasingly to data networks (X25, IPv4 or even IPv6) and potentially LAN type systems (Figure 10.14). It is important to know what presentations are available from the radio equipment and make sure the corresponding suitable interface point is available from the system multiplex or cable-trunking terminations.

10.2.6.5   Future Communication Equipment

No one can ever say for sure what form or physical shape future equipment is likely to take; however, in the interests of backward compatibility and simplicity, it is likely to be rack mountable,

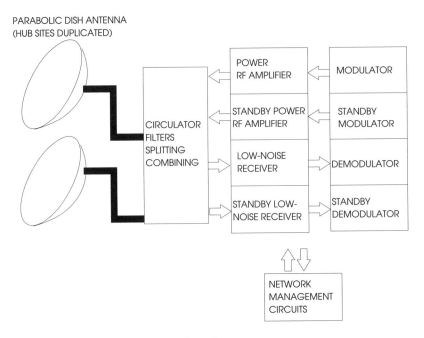

Figure 10.13   Typical VSAT equipment configuration.

powered from −24 to −48 V DC and/or 240 V AC. It will interface to existing equipment standards and connection standards and it is likely to comply with typical environmental standards. Size wise it is likely to ever decrease in size for the performance expected. For example, one anticipated future system could be based on the software or cognitive radio; this could be a typical 19-in. or ETSI mountable box, providing multiple radio functions inside one physical housing.

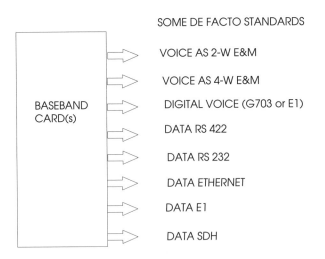

Figure 10.14   Voice/data interfacing.

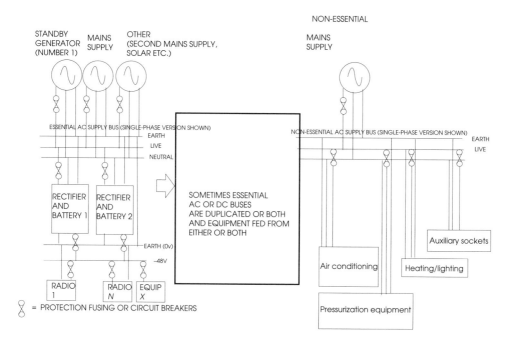

Figure 10.15 Typical power distribution layout for a typical radio repeater site.

10.2.7 Peripheral Equipment

Power equipment. Careful consideration is needed as to what power system is to be deployed within an equipment room. This is often the most common point of equipment failure which reflects how poorly considered this sometimes is by communications design and implementing engineers (Figure 10.15).

10.2.7.1 Mains/AC Service

In the first instance, it is important to understand the security and reliability of the primary mains source or high voltage (HV) source to a site: Is it on a HV ring? Is it a rural site with backup generator or uninterruptible power supply (UPS) battery inversion system (solar panels or wind generators in some cases)? In most well-engineered sites, there are often at least two primary sources of AC. (These can often even be triplicated for extremely vulnerable or strategic sites.)

Then, it is useful to understand how this is apportioned: Quite often AC is fed onto an 'essential' services bus and a 'non-essential' services bus. In theory the primary radio systems (and duplicate systems) are usually fed from the former whereas auxiliary power outlets for example external security lighting are fed from the latter. Sometimes the heating, ventilation and air-conditioning circuits would go on the latter as well except when conditioning is vital in which case they could be given a separate bus.

It is usually useful to obtain a schematic for the AC distribution for a site, or a sketch out of its configuration. (By way of recommendation here, it should be stressed that one should never assume it is as per the diagram, or fuses do not necessarily grade or always trip in the planned sequence; in cases where reliability is paramount, it is usually worth checking through the schematic in detail and, where necessary, taking advice from an independent electrical or services engineer who specializes in this.) Another useful thing to understand is the typical rating of the incoming supply and the maximum demand of all the services supplied by it in kilo volt ampere (KVA) rating.

10.2.7.2   DC Supplies

Historically, most radio equipment is supplied at DC, either 24 V, the legacy supply for mobile radio, or increasingly −48 V. The advantage of this is that battery banks can be placed in parallel to take over for 'short' periods when the AC power supply is cut or needs maintenance.

It is important to size the rectifier capacity and the associated battery ampere-hours capacity to allow for a 'reasonable' power outage window. To get the ampere hours, one should take the time in hours as the maximum time required (in poor weather conditions) to get a sizeable generator on site to make up for a shortfall incurred by loosing any AC supply. For the current draw it should be considered as the current draw taken from the essential services DC bus under steady state conditions.

Also when sizing generators and rectifiers, it should always be considered that the period of heaviest loading would be typically after a prolonged power outage when in addition to supplying the steady state equipment powers required, a substantial amount of power is required to replenish the battery bank. This can take many hours and the limiting factor here will be AC capacity and rectifier capacity.

10.2.7.3   Heating Ventilation, Air Conditioning

If (a) an equipment room dimensions are specified and thus air volumes and flows are known, (b) the typical power dissipation expected under steady state conditions for a fully fitted out (with radio equipment) room and maximum power dissipation are known, also if (c) the external ambient temperature swings, air humidity, the heat flow characteristics of the wall-insulating materials and ventilation are known, it is possible to derive a thermodynamic heat flow equation and to dimension the heating, air conditioning and ventilation requirements to keep the room well within the ambient window (permissible temperature swing and permissible humidity swing) with a reasonable degree of certainty.

In addition, it is possible to keep the ambient temperature and humidity almost fixed (usually 20 °C at 20 % RH to make a room comfortable to work in), and then the problem, deviating significantly from these, is a function of time. Again for critical applications and complex sites, it is worth engaging an expert in heating ventilation and air conditioning as to what is the recommended equipment required to maintain this environment.

10.2.7.4   Pressurization

Sometimes an afterthought is the requirement to keep waveguides pressurized. This is to keep them 'dry'. In the presence of moisture waveguide, insides can rust or corrode, which in time provides an ideal platform for intermodulation problems and thus a degraded system. Pressurization equipment is usually only required for higher frequency radio systems

(>1.5 GHz) such as microwave point-to-point systems or satellite RF components. It is important to get and keep waveguide pressurized at all stages of delivery, installation and operation.

10.3 Outdoor

10.3.1 Transmission Lines (VHF, L Band and Microwave)

With transmission lines it is appropriate to keep these as low-loss, low-mass/volume, low-cost, low-wind-loading and as easy to install as possible. For VHF this usually means low-loss coaxial cable. For L band at 1.4 GHz it can also be coaxial cable, although this can have relatively high loss and is only suitable for short runs. Alternatively the large waveguide feeder products, typically of the order of 100×60 mm elliptical cross section, can be used where this is critical. For the higher frequencies (4 and 5 GHz), only waveguide is suitable and it is slightly smaller than this. Waveguide is restrictive in going round corners, and the minimum bending radii guides must be followed. The frequencies from 1 GHz up to about 4 GHz are the most bulky. In addition cables and waveguide should be earthed at top, bottom and in bends in towers if possible and should avoid sharp bends that are likely to be attractive for lightning arcing in adverse weather.

For higher frequency systems (say 10 GHz and above) even waveguide, although in much smaller cross section (typically elliptical 40×20 mm or less), starts to add relatively high loss (typically >2 dB for 100 m), and waveguide runs must be absolutely minimized. In some cases, the active RF electronics can be moved outside to the rear of the antenna in a weatherproof housing and an IF cable run is used to connect (usually 70 or 140 MHz over coaxial cable sometimes with integral power on the coaxial cable or in separate conductors) to the indoor baseband modulator/demodulator at lower frequency (Figure 10.16).

Of course when considering a link budget, to optimize this the active transmitter and receiver should be as close as possible to the antenna. This maximizes fade margins and also provides better overall noise figure or noise temperatures, which can be very critical for some of the satellite type applications.

For a more in-depth discussion of cable guides the reader is recommended to read the Andrew catalogue (see www.Andrew.com) or the RFS catalogue (see www.RFS.com).

10.3.2 Antenna Engineering

When designing antenna systems, a number of fundamental high-level aspects need to be sorted out first. Broadly, a suitable location on a tower needs to be chosen, and consideration is also needed for what application(s) are to be used as well as for alignment and maintenance aspects and design requirements.

10.3.2.1 Antenna Location and Application

When specifying this, fundamentally the following questions need to be asked:

- Is the location to give maximum coverage or a specific (directional) coverage?
- Where will Tx and Rx electronics be, and is the feeder run optimized?
- Is line of sight required? (For satellite installations this is could be azimuth and elevation look angle.)

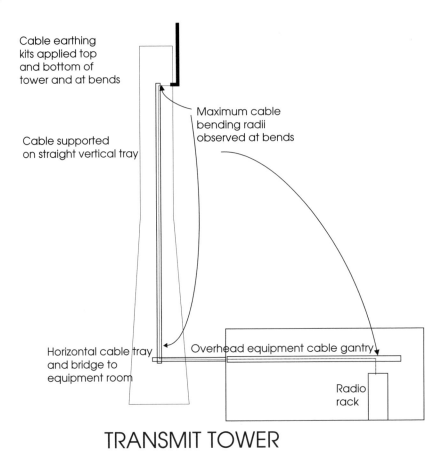

Figure 10.16 Example cable runs.

- What is the minimum height specification (from path surveying, lamp tests), obstacle clearance, multipath issues? (Is there a maximum height restriction?)
- How is the antenna going to be operated (stand-alone, with space/frequency diversity, climax, etc.)?
- Can other antennas or metal work on a tower cause problems to the antenna directivity or increase risk of swamping or intermodulation aspects, blocking Fresnel zones, etc.?
- Tower constraints. Sometimes tower owners price antenna locations as a function of height (and size for dish or solid antenna types), this is a reflection of the civil structure cost which is function of tower loading.
- Are there significant wind loading or ice loading (plus counter measures) considerations to be made?
- What are the consequential feeder runs expected, how are these protected (are they minimized from a design perspective)?
- Accessibility. Can the equipment be easily reached from working platforms or not, in all weathers? What climbing certificates will be required for working on equipment, how

'easy' will the equipment be to maintain? What are the typical weather restrictions, also accessibility or restrictions are in place to prevent public or third parties getting to equipment (e.g. protection of waveguides), and the reciprocal of this (the protection of third parties and public from hazards from the equipment such as RF, falling ice, etc.)?

• What RF safety issues need consideration? (This could be for working staff and the public.)
• What about environmental's? Are council approvals required or site approvals?
• What labelling convention is required?
• What polarity of the equipment should be operated?
• Can the installation be 'standardized' with antenna types and feeder types, etc.? (This is to minimize spares holding and rigging training.)
• Measures to protect from exposure to wind, ice, rain, sun, etc.
• Is duplication or diversity required?
• What about co-site aspects?

10.3.2.2 Antenna Selection

Hand in hand with the antenna location and application process is the antenna selection. This is where considering engineering requirements, the optimum antenna can be selected. To do this the following consideration should be given.

Directional/non-directional. That is, how directional an antenna is required, is it for coverage, is it for range or is it for a specific service volume or quadrant, where is the radio tower location in relationship to this service volume? Or is a directional antenna required for a point-to-point link?

Gain. What is the centre beam gain of this antenna in dBi (or dBd)? What is required in the link budget equation?

Radiation pattern envelope elevation (RPE). What is this in azimuth pattern and elevation? What is the front-to-back ratio or F/B? (This is the relative difference in gain of the front of the antenna versus direct 180° rear.) Also *beamwidth* can be another dimension to the RPE pattern; it is defined as the points on the Cartesian diagram where the directional gain falls off at 3 dB from boresight. (The RPE and these related parameters of beamwidth and F/B are important parameters when considering coordination of frequencies and co-channel and adjacent channel interference aspects; further discussed in Chapter 12).

Bandwidth. What is the RF frequency bandwidth of the antenna in terms of 3-dB points? Is a wide band or narrow band antenna preferable?

Impedance. This is the electrical impedance of a matched antenna. Usually radio systems are matched for 75- or 50-Ω working. A related part of the antenna specification will also be its VSWR, which is the degree to which an antenna and its transmission line can tolerate mismatch and will describe the ratio of the antenna return reflected signal to the absorbed signal input into an antenna. Mismatch of antennas to transmission lines and Tx or Rx will create errors to the information being passed and in the ultimate extreme will render a system unusable.

Polarization. This is an important consideration for licensing. Related to this is the *cross-polar discrimination* (sometimes called *XPD*) of an antenna which shows the level of rejection an antenna has to a plane polarized signal coming in the alternative field of excitation.

(With mobile communications generally only vertical polarization is used, for point-to-point links vertical and horizontal polarization is used, for satellite communications, circular polarization is used where the polarity changes with each wavelength either in a right hand corkscrew motion or in a left hand direction.

Power handling. Usually the power-handling capabilities of an antenna need considering to ensure the antenna is of suitable rating.

Connexion. A multitude of connexions are available from antennas. When interfacing to coaxial cable usually F-type connectors, BNC connectors, *N* type connectors and SMA connectors are used as the industry standard. When interfacing to waveguide the connector is usually a rectagonal window (that is usually pressurized with a mica seal at some points in the system). The dimensions of the window are a function of the waveguide being used. Again the reader is referred to the Andrew, Gabriel or RFS catalogue. For active power or data interfaces to external electronics, there are hundreds of proprietary connector types and are too numerous for this book; they are usually equipment specific, and it is suggested to look on the equipment technical specification brochure. Also sometimes sniffers are available on the back of active antennas where antennas can be aligned without breaking principle feeder chain, or even active AGC voltages can be found from head-end mount electronic devices.

Mounting arrangement. Once an antenna is selected, the all-important hardware for fixing it to the mast or concrete plinth needs to be made. Care is required when selecting these to reconsider all of the points raised above.

10.3.2.3   Alignment and Optimization

When antennas are installed it is important to align them and optimize them. For directional antennas, alignment usually involves measuring an active source from the far end (and using a power metre or spectrum analyser) and peaking the antenna. For omnidirectional antennas, the reverse is usually carried out; it is installed and a drive or flight test is used to check the all round coverage capabilities of the installation.

Also equally important is to make sure the antenna is properly matched to the transmission line and the various connectors through to the receiver system. This can be done by sweeping the antenna and components (individually or collectively). This is when a wave generator is sent into the antenna system though a directional coupler and looks at the return echo given by mismatch VSWR or band connections (Figure 10.17).

10.3.2.4   Practical Antennas

A number of practical antennas are listed below.

Example omnidirectional antenna (suitable for ATC VHF communication) (Figure 10.18)

Rohde and Schwarz.
Bandwidth: wide band frequency range 100–156 MHz;
Physically: self-supporting antenna mast, VHF dipole, rugged;
Gain: 2 dBi gain (per dipole, multiple dipoles in housing)
VSWR: <2;
Power rating: 1 kW max;
Temperature range: −40 to +85 °C;

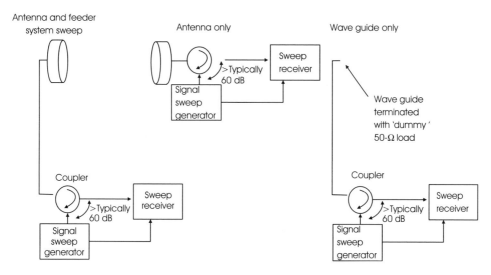

Figure 10.17 Antenna 'sweeping' VSWR conformation.

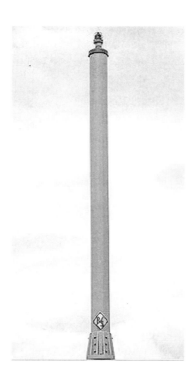

Figure 10.18 Omnidirectional antenna (suitable for ATC VHF communication).

Figure 10.19   Example omnidirectional dipole (Andrew DB224-FAA).

   Polarization: vertical;
   RPE: (azimuth) circular pattern.

Example omnidirectional dipole (Figure 10.19)

   ANDREW DB224-FAA;
   Omnidirectional, exposed dipole;
   Frequency range: 127–141 MHz;
   Gain: 6/8.1 dBd/dBi (4 stack dipole collinear array);
   Horizontal BW: 360°;
   Vertical BW: 16°;
   Polarization: vertical;
   Beam tilt;
   VSWR: <1.5:1;
   Size: 7087 × 44 mm;
   Wind load: 560 N;
   Connector: *N*-type male (1) bottom.

Example omnidirectional antenna (Figure 10.20)

   Rohde and Schwarz. HK012;
   Frequency range: 100–160 MHz;
   Omnidirectional VHF antenna with vertical polarization, featuring high suppression of skin
      currents;

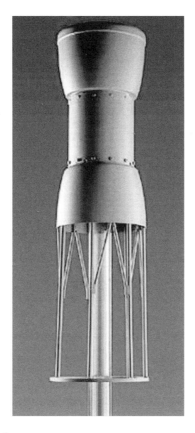

Figure 10.20 Omnidirectional antenna.

Gain: 2 dBi;
Impedance: 50 Ω;
VSWR: <2;
Power handling: 100-W max;
Temperature range: −40 to +85 °C;
Vertical polarization;
Circular pattern;
Rugged;
Wide band frequency range.

Broadband omnidirectional antenna (Figure 10.21)

Rohde and Schwarz HK014;
Broadband 100–1300 MHz (extendable from 80 to 1600 MHz);
Extremely broadband omnidirectional antenna with vertical polarization featuring high suppression of skin currents;
Low weight;
Minimal wind load;
Sturdy design;

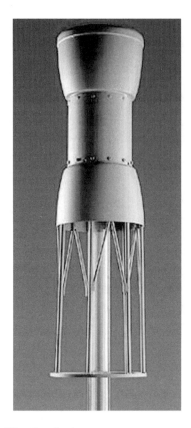

Figure 10.21 Broadband omnidirectional antenna.

Protected against lightning;
Vertical pattern with null fill-in;
Impedance: 50 Ω;
VSWR: <2;
Maximum power: 1 KW;
Gain: 2 dBi.

Log periodic test antenna (Figure 10.22)

Broad frequency range: 30–1300 MHz;
Dual polarization;
Typical gain: 6 dBi (varies with frequency);
VSWR: 2.5.

Parabolic dish antenna (and satellite antenna) (solid, high-performance (with shroud) type)
(Figure 10.23)

High performance (i.e. additional attenuation applied to off-beam components).
These are plane polarized (or can be dual polarized dual feed).
The gain is a function of antenna size and frequency.

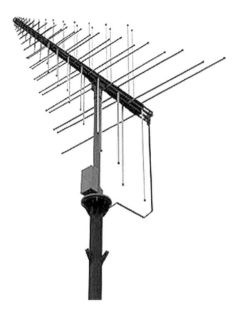

Figure 10.22 Log periodic test antenna.

RPE is also a function of antenna diameter (type geometry) and frequency.

Mounting arrangement is usually an A or inverted-A frame (see manufacturers specifications).

Radomes if used usually low-loss Teflon with loss of less than 0.5 dB.

Low wind loading and Gridpack variants allow similar slightly reduced electrical specification with lower wind-loading properties.

Figure 10.23 Parabolic dish antenna (high-performance, low-wind-loading and Gridpack versions).

Figure 10.24 High performance parabolic antenna (see Shrouds). Reproduced from Gabriel Electronics.

The satellite ground station antenna is usually a variant on this (Figure 10.24). (Usually just much bigger; hence why it is usually ground mounted on a plinth.)

HF ground antenna (Figure 10.25). Usually for the MWARs an omnidirectional antenna is used.

Vertical polarization;
VSWR 2.0:1;
1.6–32 MHz (wideband).

10.3.3 Towers or Masts

Again a speciality that this book does not intend to go into this topic in any great depth; for this it is recommended that appropriate civil or structural engineering documentation is consulted, or even better consultancy services from a specialized and reputable engineer or manufacturer.

At the heart of tower engineering, towers are designed for purpose. This may be initiated by tabulating a list of wanted antenna types and location requirements for a new tower. Some features can be pre-empted in this process (Table 10.1).

At this stage this basic 'requirement specification' can be submitted to a tower specialist, who in turn usually has a computer-based program (based on finite element analysis or basic stress/strain analysis) and they can 'crunch' a tower design out. Normally the tower design will give an all-important 'loading figure' on as a percentage of full capacity. (This is a function of the civil and structural engineering codes in place in the country under discussion and the statistical site wind-loading factors and can vary from site to site and country to country.)

From past experience, the optimal design or a good benchmark is for a tower loading around 75–80 %. Above this and there is very little margin for safely increasing the loading on a tower without reassessment and re-engineering and sometimes 'stiffening' of the tower sections.

Figure 10.25 HF omnidirectional antenna. Reproduced from Andrew Corporation.

Also in the bigger picture it means the tower is ultimately more sensitive to any big freakish storms that could blow it over or deform it. In contrast, lesser loadings indicate an unnecessary conservatism when specifying a tower, or it has a cost associated with it.

If substantial modifications and/or additions are required to a tower, it is usual practice to rerun through the tower loading analysis, which is usually quite a quick business and can tell the necessary upgrading requirements for a tower, if any.

10.3.4 Equipment Room

Equipment room design is another specialization combining the building engineering and services skills with the end user (or radio engineering) skill set. Equipment rooms can be of permanent (brick or concrete) construction or more temporary (e.g. sandwich wall construction),

Table 10.1 Example tower requirements.

Antenna number	Antenna height Above GL/location	Antenna type	Orientation	User
1	24 m/SE leg	Omni directional (Tx only)	Omni (coverage to sector)	Aviation authority
2	12 m S face	2.4 m Dish HP	162°	Aviation authority
3	18 m NW leg	Φ3 m Dish gridpack (diversity)	234°	Electricity
4	38 m NW leg	3.7 m Dish gridpack (main)	234°	Electricity
5	40 m (top) top	Omni (Tx only)	Omni	Aviation authority

Figure 10.26   Example control tower with VHF antennas on the roof.

depending on specific application and environment and increasing budget. Most layouts are common sense, optimizing available space and suite management against minimizing feeder runs and routes and optimizing the engineering link budget designs required. The detailed principles are usually documented by site owners (such as civil aviation authorities or ANSP codes of practice) in the form of engineering practice documents.

When first setting up an equipment room it is important to plan it properly. The following principles/tick list should be applied.

1. What is the most efficient use of the available floorspace?
2. Is front and back access to racks required? (If in doubt assume yes.)
3. Adequate gangways (if in doubt 600 mm to 1 m is usually good practice).
4. Where will the transit be for the cables?
5. What are the overhead ladder racking plans?
6. Power supplies and power distribution board (both for AC and DC).
7. Interface points to RF or baseband equipment.
8. Connections required to multiplexing, trunking or cable terminations.
9. Is there a cable termination patch panel?
10. Is there a space reserved for third-party service providers or PTT equipment?
11. Is pressurization required?
12. Maximum bending radii considerations of cables.
13. Height of room and consequently ladder racking.
14. Location of mains (non-essential supply) outlets.
15. Lighting.
16. Heating, ventilation, air conditioning.
17. Door access and size: Can equipment easily be brought in and out?
18. Telephone access.
19. Is there plans for network management of the equipment?

Beyond this, high-level view is considered outside the scope of this book.

10.3.5 Equipment Racks

There are many standard equipment racks available which will house aeronautical equipment. The industry tends to standardize on the following type systems:

- 19-in. rack: This is nominally 19-in. (482.6 mm) internal to the rack, outside walls of the rack can range from 500 to 600 mm.
- ETSI racks: These are 600-mm wide × 300-mm deep (two can fit in the footprint of a 19-in. rack).
- Equipment supplier proprietary racks (e.g. Nortel radio): These footprints are defined in Nortels NTPs but are a variation on a 19-in. rack.
- Heights of racks vary from a half rack starting at 1.0 m to typically 1.8 or 2.0 m and occasionally 2.25 and 2.50 m.

For further detail see equipment manufacturers' technical specifications or specialists in equipment racking.

11 Avionics

Summary

This chapter deals with the practicalities of realizing radio equipment on-board an aircraft. It starts by looking at the equipment environment, then it looks at all the component parts of a radio system. The avionics modules are usually installed in the avionics bay or in smaller aircraft in the cockpit, the cockpit-mounted control modules and avionics, the antenna and cable systems and ancillary equipment.

It concentrates on the unique environment with severe constraints on space, power draw and temperature dissipation and extreme environmental swings. It is an interesting contrast to the previous chapter where the radios are system-wise almost the same but the environment and scenario are totally inverted.

It also relates the avionic modules to the various compliance standards, testing standards and certification requirements.

11.1 Introduction

The environment on-board or on the external skin of an aircraft is in stark contrast to anything terrestrial. It has been described in radio terms as a $>10\,\text{dB}$ increase to everything terrestrial. This is not quite as exaggerated as it seems. In terms of wind loading on the skin of the aircraft, wind speeds of $>1000\,\text{km/h}$ are steady state for an aircraft such as a commercial jet (and even more for specialized supersonic aircraft) for typically more than 50 % of the aircraft's working life. Similarly temperature cycles between $+50^\circ$ C down to -50° C (or even more) three or four times a day can be a reality of everyday operations. Severe vibrations can be expected at least a few times a day. During all this, high performance and reliability is expected from the radio system at all times, and this can be critical to the safety of flight.

11.2 Environment

Table 11.1 summarizes the high-level environmental conditions on an aircraft. This is usually the extreme case. In some cases the extremes may not be reached; for example, consider a

Aeronautical Radio Communication Systems and Networks   D. Stacey
© 2008 John Wiley & Sons, Ltd

Table 11.1 Aircraft environmental conditions: an overview.

	Inside	Inside non-pressurized	Outside
Temperature	−55 to +85° C	−55 to +85 C	−55 to +85° C
Humidity	0–99 %	0–99 %	0–100 %
Pressure	0–1.697 atm	0–1.697 atm	0–1.697 atm
Gravity forces	Variable, typically up to 9 Gs impulse operational	Variable, typically up to 9 Gs impulse operational	Variable, typically up to 9 Gs impulse operational
Other weather conditions (salt fog, sand, dust, fungus)	Various	Various	Various
Power	28 V DC, 115/200 V AC	28 V DC, 115/200 V AC	28 V DC
Weight	Minimal	Minimal	Minimal and aerodynamic
Size	Minimal	Minimal	Minimal and aerodynamic
Standardization	Yes to rack specs	Yes to rack specs	Yes to skin penetrations and specifications
Certification	Yes, mandatory	Yes, mandatory	Yes, mandatory
Vibration/impact	Yes, military specification	Yes, military specification	Yes, military specification
Weather protection	Lightning, static	Lightning, static, pressure, temperature	UV, static, lightning, pressure, temperature, rain, icing
Safety	Detailed cooling, ventilation specification	Detailed cooling, ventilation specification	
Reliability	Various specifications	Various specifications	Various specifications
Connections	Various industry standards	Various industry standards	Various industry standards

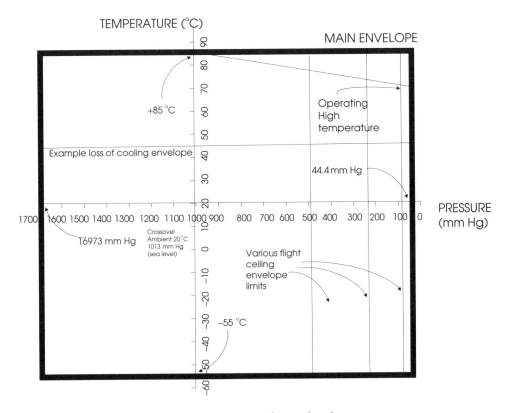

Figure 11.1 Graph with temperature versus pressure bars and equipment.

non-pressurized aircraft with a ceiling of 10 000 ft where the outside temperature would not be expected to reach –50° C) (Figure 11.1).

In the following, we shall be looking at the environmental considerations in more detail.

11.2.1 Temperature

11.2.1.1 Outside

The outside temperature range of +85° C down to −50° C is very realistic for an aircraft working in the Middle East or tropics. On a hot day, the aircraft skin is in full view of the sun and can actually heat significantly above the ambient temperature of the air. Add to this the heat dissipation needed by the aircraft and the fact that some of the outside antennas can be active with their own residual heating. This gives a worst-case exterior temperature of typically +85° C. Conversely, when an aircraft reaches cruising altitude between 30 000 and 40 000 ft, the outside air temperature with wind chill is almost a constant −50° C or worst case −55° C.

Associated with the temperature swing can be a significant expansion/contraction, particularly of metallic parts. (For example, Concorde was quoted to show a 6-in. difference between its ground length and its flight ceiling length.) This in turn means that cabling lombs, connectors and electronic component boards are moving and there is a change in stress and strain as they are temperature cycled.

11.2.1.2 Interior

Temperature swings are less radical in the interior than on the aircraft exterior (except, maybe, for the electrical power bay where ambient temperature is usually higher than outside) but can still swing appreciably.

11.2.2 Pressure

11.2.2.1 External Pressure

Typically at the ground level the pressure is nominally 1 atm, and at an altitude it can be less than a quarter of this. Design specifications typically call for overpressure values of +1.697 atmospheres and down to 0.044 atmospheres (or approximately 0).

11.2.2.2 Internal Pressure

The indoor environment is generally less harsh under steady-state conditions; however, rate of change of pressure differences can be more drastic as compared to external pressure, particularly when there is unexpected loss of pressurization. It can change from 1 to 0.25 atm in a few seconds.

Distinction should be made between pressurized and non-pressurized sections (see next section on testing). Obviously, under the pressurized sections the pressure and temperature swings should be less radical; however, under some unforeseen operational conditions, cabins, cockpits and avionic bays do become decompressed, under which conditions the equipment is very much required to be fully functional and the specification remains much the same as for non-pressurized areas.

11.2.3 Equipment Testing

Taking into account the general discussion on environmental conditions above, it is necessary to specify limits and test equipment to these accordingly: The RTCA Document DO 160E 'Environmental Conditions and Test procedures for Airborne Equipment' goes into much more detail on this and classifies aeronautical environment into a number of categories for temperature and pressure (Table 11.2), which are as below.

For this there are much more detailed criteria for each of the categories; for example, temperature can be broken down into operating low temperature, operating high temperature, short-term operating low and high temperatures, loss of cooling test temperature, ground survival high and low temperatures, with variable pressure, overpressure or decompressing parameters. This book does not deal with these in-depth, but the extremes are depicted in Table 11.1; if more detail is sought, see RTCA DO 160E.

11.2.4 Apparent Wind Speed

At cruising altitude a typical jet is flying just below Mach 1, with a typical cruising speed of 1000 km/h and maximum velocity of 1200 km/h (air speed, not ground speed). Wind rating: typically 1200 km/h for commercial jet (faster for military jets so take on a case-by-case basis, slower for General Aviation or Turbo Prop operations).

Table 11.2 Environmental categories according to RTCA DO 160E.

Category	1	2	3	4
A	Controlled temperature, non-pressurized to 15 000 ft or pressure controlled 0–15 000-ft equivalent pressure	Partially controlled temperature, non-pressurized to 15 000 ft or pressure controlled 0–15 000-ft equivalent pressure	Controlled or partially controlled temperature but are more sever than in 1 or 2, non-pressurized to 15 000 ft or pressure controlled 0–15 000-ft equivalent pressure	Ditto as 1, except that temperature conditions differ from category A1 as declared by equipment manufacturer
B	Controlled temperature but non-pressurized up to 25 000 ft	Non-controlled temperature, non-pressurized up to 25 000 ft	Equipment that is operated in the power plant compartment of aircraft, at altitudes up to 25 000 ft	Non-pressurized up to 25 000 ft, temperature requirements different from B1 and B2
C	Controlled temperature, non-pressurized up to 35 000 ft	Non-pressurized, non-controlled temperature up to 25 000 ft	Equipment that is operated in the power plant compartment of aircraft, at altitudes up to 35 000 ft	Non-pressurized up to 35 000 ft, temperature requirements different from C1 and C2
D	Controlled temperature, non-pressurized to 50 000 ft	Non-pressurized, non-controlled temperature up to 50 000 ft	Equipment that is operated in the power plant compartment of aircraft, at altitudes up to 50 000 ft	No such category
E	Non-pressurized, non-controlled temperature up to 70 000 ft	Equipment that is operated in the power plant compartment of aircraft, at altitudes up to 70 000 ft	No such category	No such category
F	Controlled temperature, non-pressurized to 55 000 ft	Non-pressurized, non-controlled temperature up to 55 000 ft	Equipment that is operated in the power plant compartment of aircraft, at altitudes up to 55 000 ft	No such category

11.2.5   Humidity: 0–100 %

11.2.5.1   External

Obviously when in the desert or at the poles, the humidity will be close to 0 %, in the tropics it approaches 100 % and when it rains it is 100 %. Consideration should be given to driving water conditions and seals or encapsulation to prevent penetration.

11.2.5.2   Internal

Humidity is unlikely to reach 100 % unless external doors are left open, but it is known to approach this level inside. In some ways this can become even more problematic as once condensing has started inside the aircraft it is harder to get rid of humidity externally to the aircraft.

11.2.5.3   General

In RTCA DO 160E, three categories are defined: category A – the standard controlled humidity environment; category B –'severe humidity environment'; and category C– 'external humidity environment'. The proposed testing for all equipment cycles the equipment through an even test. This gives a maximum humidity of 95 % ±4 (so 99 % worst case) with varying temperature. This test is specifically designed to cause condensation in the equipment and thus test for this.

   Waterproofness is a concept presented in RTCA DO 160E There are a number of increasing levels of this:

Category Y, internal where equipment is subject to condensing water in course of normal operations.
Category W, where equipment is subject to running or flowing water, usually as drain for condensation.
Category R, where equipment is subject to driving rain, or water spray from any angle (test from showerhead 2.5 m from equipment, shower at 450 L/h, minimum 15 minutes)
Category S, where equipment is subject to forces of a heavy stream of fluid as encountered under de-icing or washing (spray test through 6.4-mm nozzle a jet that can rise 6-m vertically, from 1 or 2 m).

   Specifications for *penetration of seals* are dealt with in MIL-STD-810E tests.
   Ice loading. This is possible particularly on the under-fuselage antennas when standing and under certain airborne conditions, though not on wings except maybe on taxiing as these are usually sprayed with a de-icer before take-off in adverse conditions. Three categories are described in RTCA DO 160E:

Category A is applicable to externally mounted equipment, non-temperature controlled.
Category B is applicable to equipment with moving parts.
Category C is where there is risk of ice build-up.

   Real estate. Externally, this is at a premium. Real estate is limited to the skin of the aircraft, and antennas of minimal wind loading are preferred. In radio terms, antenna separation distances cannot always be optimal. It is a compromise between space available and electrical RF isolation between antennas; this brings co-site radio antenna issues (self-interference), which get complicated very quickly.

Weight. As per all avionics, minimizing weight is critical and eventually equates to less fuel burn. Externally, weight is not as important as the wind loading and aerodynamics, however, is worth investing in super lightweight components and encasing. Indoors, weight is also at a premium and is to be minimized.

Also weight has an important relationship to gravitational forces, and as per elementary physics this is the distinction between weight and mass: weight is mass multiplied by gravitational forces. So in a severe manoeuvre or a crash situation, the weight forces increase with G, up to 10 or even 20 times; hence the real reason why weight needs to be minimized.

Size. More important to weight on the external surface of an aircraft is the size, or the wind drag forces associated with this and aerodynamics (as these are more significant that just the weight). Emphasis is made on minimizing the drag and making antennas as aerodynamic as possible. Again, suitable materials are chosen for this purpose. The concept of a 'blade' antenna with minimal forward surface area and moulded flat active satellite antennas is favourable over the dish type antennas.

Size is generally not as critical internally as the external mounting equipment; however, equipment size or more accurately space (as the equipment is installed in modular units) must be minimized. It is also worth noting that designers generally prefer to compromise on weight and size in the interests of the modular line replaceable unit concept. It is a fine trade-off, which in the future will have an increasing importance particularly as technology enables smaller size and weight, and fuel prices potentially move the fulcrum for weight.

Gravity and forces. Under conventional manoeuvring, a jet can get up to 2-G forces on it. Military jets in practice use much more than this and potentially in the new generation of unmanned aerial vehicles they can exceed the human endurances. However, more significant are the vibrational short-term impulse impacts, which can go as high as 9 Gs under extreme operational environments; above this there are crash impacts for which survival tests have been defined up to 20 Gs.

Explosive atmospheres. By its very nature conventional aircraft are carrying fuel, and equipment is exposed or in close proximity to the fuel and fuel systems, purge lines, vents and filling recepticals. From RTCA DO 160E, various environment categories are specified (usually depending on the degree of proximity to this hazard and the degree to what a device can cause heating or a spark):

Environment I is the most extreme category where uncovered flammable fluids or vapours exist.
Environment II is where flammable fluids or vapours can exist but only after fault or accidentally after spilling.
Environment III is within a designated fire zone.

Corresponding equipment specifications are also defined:

Category A is for hermetically sealed, i.e. any fire or flash is self-contained within the seals.
Category E is for non-hermetically sealed equipment; it cannot usually be used in environment I.
Category H equipment contain hot-spot surfaces internal and/or external but are non-spark producing under normal operating conditions. Under normal operating conditions the external surfaces should not rise to a level capable of causing ignition.

Weather protection (fluid susceptibility, sand and dust, UV, salt fog, dust, fungus). There are various specifications to prevent the ingress of or damage from these weather conditions. To prevent against these and weather aspects described previously, external equipment should be ruggedized, i.e. its tolerance to this harsh environment, by, for example, encasing antennas in immune toughened smooth plastic.

Fluid susceptibility is an aspect applicable to areas where fluid contamination can take place (e.g. near engines, fuel lines, lubrication points, hydraulics).Necessary countermeasures are described in RTCA DO 160E.

Sand and dust when carried by air can cause havoc with joints, cracks and crevices: In that it can start to add to or accelerate an abrasion process or it can form conductive or semiconductive bridges on circuit boards; it can pollute fluids, leading to the fluid susceptibility problem and can be corrosive. Two categories are defined in RTCA DO 160E for this: category D and category S, both are for equipment installed where there is susceptibility to dust and sand, the latter being used when equipment may have moving parts. Also abrasion of surfaces and erosion of surfaces from such particles and specific tests is defined in MIL-STD-810E.

Fungus growth. In some regions of the world, it can be common, fuelled by environment, UV, high humidity and nutrients. Fungal growth can cause adverse effects to certain equipment types whose structure could be broken down with this in time. There are categories for equipment installed in such environments as well as tests to confirm equipment.

Salt fog or excessive salt materials cause a pH imbalance that can cause corrosive effects to certain materials, leading to insulation problems, clogging of moving parts or bad mating of connections/contacts. In an antenna system, they can also lead to the secondary effects of intermodulation. Two equipment categories are defined in RTCA DO 160E: one for normal operational exposure to salty environments (category S), other for extreme conditions when salt spray is incumbent under routine conditions on equipment (category T).

Power. Power to the outside of the aircraft skin is only to active antennas (such as satellite antennas) usually in the form of 28 V DC.

Electrical connections. See MIL-STD-810E tests that define the quality of electrical connections, mating of components and degradation limits of electrical circuits.

Standardization. Emphasis is given to standardize the antenna types and fixations on-board an aircraft so that there is flexibility to adapt these, as operational missions may change.

Certification. For flight operations, all commercial aircraft in the country have to be certified by aircraft registration (usually by the appropriate Civil Aviation Authority, CAA). As part of this process the proper installation and ongoing integrity of external antennas needs to be regularly confirmed.

Vibration/impact. All externally mounted equipment needs to be designed for continuous vibration throughout the life cycle: that is, for shock impacts mainly from take-off run, touchdown, landing run-out and general turbulence; in addition, vibration can be experienced from engine imbalances. One of the main problems is secondary collision, when on the airport apron, by service vehicles loading and unloading the aircraft.

Vibration is also a function of how far away equipment is from the centre point of the aircraft about which things vibrate. Also local elements such as wing tips or tail fins have local resonating frequencies, and installations around these points are more prone to vibration.

Vibration conditions are much the same for aircraft internal equipment.

For vibration there are a number of specifications.

- Standard vibration test (described as category S in RTCA DO 160E). This is for the standard operational environment experienced by fixed wing aircraft.

- Robust vibration test (described as categories R, U, U2 in RTCA DO 160E). This is a prolonged overvibration test for fixed wing and rotor aircraft.
- High-level short duration vibration test (described as categories H and Z in RTCA DO 160E).

For impact shock, four categories are defined in RTCA DO 160E:

Category A is for equipment designed for standard operational shocks.
Category B is as for A, additionally tested for crash safety.
Category D is equipment tested for low-frequency shock.
Category E is equipment tested for operational low-frequency shock and crash safety.

The tests include random forces applied to equipment and fixings, typically 6–9 *G*s (gravitational force equivalents) of force for normal operation and up to 20 *G*s for some of the random design crash forces. The crash safety is the most stringent vibration force and the emphasis here is on survival of equipment and tolerance of fixings and holdings.

The location of equipment on-board an aircraft also defines what vibration and impact specifications are relevant. For example, vibration is more severe in close proximity to engines or at extremities of aircraft far from the centre point 'fulcrum' and also at wing and tail tips. Crash impact survival is mainly pertinent to areas around the cabin and cockpit, where it is more essential that equipment does not break from its mounts and become a lethal projectile.

Safety, fire resistant, fire proof, flammability. All external equipment needs to be 'safe' for the environment, that is, for ground and air safety. Safety includes aspects such as field strengths (flashpoint voltages) in the proximity to aviation fuel, RF personnel hazards and physical hazards of the blade shapes. Internally, the equipment needs to be designed so that it is fail-safe, fire resistant, smoke proof and is usually constructed from safe low-smoke and low-fume type materials, wherein the threat of generating fire, excessive heat or smoke is minimized.

Equipment is generally designed to have some resistance to fire, either fire proof or fire resistant. The details of this are described in RTCA DO 160E, Section 26.

Connectors. Most antennas are standard bolt on to the fuselage skin, with suitable seals between the antenna and aircraft. Cables are usually internal to this and usually use standard (F type) connectors, embedded inside the antenna casing. Specifications and code of practice are aircraft type and equipment specific.

Power, electrical transient protection, lightning and electrostatic discharge. This is an interesting and complex aspect for internal equipment. It has a double whammy effect. Firstly, power draw must be minimized as AC or DC power generation is at a premium on-board an aircraft. Secondly, in addition and even more crucial, as with most electronic components, most of the power drawn is dissipated in the form of heat to circuit boards and modules, which is definitely an unwanted commodity, and electronic design is geared to minimizing this current draw but more importantly, the heat dissipation from units – under some instances fan circuits are required. This, of course, in turn exacerbates and adds to the complexity of the thermodynamics and power draw. In the limit, if temperatures cannot be contained, they would become a safety issue (for ignition/sparking, etc.), and very strict specifications and codes are enforced to avoid the equipment ever getting near this point. Also much work has to look at the failure modes of avionics (for example, overcurrent and overheating or fan failure), which becomes very complex, very involved and very expensive.

The necessary protection of power systems (circuit breaker grading) is also very important. Most modern aircraft operate an active circuit breaker where the status of the aircraft breaker is reported to the flight management system over the data bus described.

Lightning effects and the transient susceptibility of equipment is described in RTCA DO 160E, Section 22 and 23.

Finally, electrostatic discharge is a danger to electronic and other components and a risk to fuel ignition. Part 25 of RTCA DO 160E describes test specification applicable to all equipment for installation in aerospace environment (basically a 15 kV impulse test).

Standardization. Avionics are standardized into modular units, which enables quick swap outs and configuration flexibility. Connectors, back planes and data standards are also largely standardized. (This is discussed later.)

Reliability. As previously described in Theory section (Chapter 2), emphasis is on equipment reliability in terms of maximizing MTBFs and also on being able to perform quick modular change-outs. Some of the avionic equipment is absolutely vital to the safety of the flight at various stages of flight.

Connectors. Most connectors are of industry standard and have been designed for the tough environment just described.

11.2.6 RF Environment, Immunity, EMC

These will be discussed in much more detail in Chapter 12. However, the aircraft by default is a highly electromagnetic and RF noisy environment. This electromagnetic activity could play havoc with the various communication system buses and RF avionics if proper measures were not put in place to protect the susceptibility of each of the devices. Much of RTCA DO 160E is devoted to quantifying this environment on an aircraft, setting protection limits for susceptibility and qualitative testing standards to prove this, in particular the following:

- Section 19 looks at induced signal susceptibility from power systems and large current power transients.
- Section 20 looks at the susceptibility of cabling avionics and RF systems to unwanted RF signals or the interaction of different systems.
- Section 21 looks at how to test for unwanted radio emissions.

11.2.7 Environmental Classification

The overall environmental classification of a device can be defined from the serial code number found on the equipment related to RTCA DO 160E (Figure 11.2).

11.3 Types of Aircraft

Before going into the detailed discussion on avionics, it is important to analyse the different types of aircraft, their associated operational and commercial driving forces and the consequent ethos and attitude to avionic requirements. Aircraft can be categorized into many different groups according to size, take-off weight, function, engine type, commercial or private, wing type, fixed, moving wing or rotor etc. – the list is endless. Each state has subtle differences

Figure 11.2 Equipment nameplate marking.

as to how aircraft is classified and the detail of this is best found in the classification register from the relevant CAA or the JAA (Joint Aviation Authority).

For purposes of discussion, however, as an oversimplification, aircraft in this book can generally be categorized into the following four types: private aircraft, general aviation, commercial aircraft or military aircraft. It should be pointed out that these are very general classifications to enable the discussion to go further.

11.3.1   Private Aircraft

These are usually hobby aircraft, flown at the weekend by enthusiasts, private pilots. (An example could be a turboprop CESNA four-seater or a Piper Warrior four-seater, or a two-seater hobby helicopter.) In general it would be fair to say the configuration of these aircraft would be budget driven. That is, an aircraft would be typically in the region of €100 000, with operating costs typically 10 % of this per year. This puts it in the hand of the executive or upper middle classes and above only or shared in a consortium.

For the application the aircraft would be 'equipment minimalist'; i.e. it would only carry the necessary mandated avionic requirements for communication and navigation rated for clear visibility or limited instrument flying conditions for use mainly in non-controlled airspace. As it is usually single-engined, only the ignition wiring would be duplicated but there would be no other duplication of communications and navigation equipment (other than maybe portable GPS and portable VHF transceivers carried on-board). Also of note is that the 'cockpit' and 'cabin' are the same thing and serve as flight deck and passenger lounge, equipment room and luggage storage (Figure 11.3).

In summary, generally private aviation can be considered to have

- no need for equipment to be interconnected via a data bus, all stand-alone modules;
- all equipment mounted in cockpit.

11.3.2   General Aviation

In contrast, there is also the private jet, e.g. for diplomat, pop star, VIP or millionaire company executive type. (An example may be a Lear 60 or a Gulfstream 100 which is in the tens of millions of euros purchase region, with higher than 10 % associated running costs.) These tend

Figure 11.3 Private aircraft.

to be customer driven, with an emphasis on comfort, speed, flexibility and are often chartered. They tend to have a segregated cockpit and cabin, with the avionics either mounted in the cockpit or in a separate avionics bay (Figure 11.4).

These aircraft are usually much more heavily utilized, particularly under charter conditions when the emphasis is on continuous operation. The avionic requirement is consequently matched and less budget restricted. Duplication of many of the essential communications and navigation functions is typical to enable its certification for instrument flying in most international categories of airspace including that controlled by ATC or sometimes military in all regions of the world. In the extreme charter category, these general aviation aircraft sometimes can be considered more like a commercial airliner.

11.3.3 Commercial Aviation

Commercial aviation is usually entirely business driven, with a simple business case based on revenue from ticketing = all purchase, leasing, operating and running costs of aircraft plus profit. This commercial emphasis has a very direct bearing on the reliability economics, with a maintenance downtime having a huge running cost implication. Consequently many of the mandatory provisions such as avionics are duplicated and even triplicated. With this requirement, real-estate space in the cockpit very quickly runs out due to space and heating effects. It is necessary to adopt an approach where avionics are located remotely as modules in an avionics bay or rack, with the control of these devices only being brought to the cockpit (via a console or controller).

This modularity concept to ease maintenance and troubleshooting becomes increasingly important for this application. Also there is a move to distributed functions for reliability (e.g. avionics bay separate from cockpit, distributed communication buses and sometimes duplicated avionic bays and power supplies.). There is also highly computerized equipment monitoring

Figure 11.4 General aviation aircraft.

from the aircraft flight management system all equipment connected over duplicated common buses with built-in fault diagnostics (Figure 11.5).

In summary, generally commercial aviation can be considered to have

- a need for a distributed standardized communications environment with common interface and interconnection between equipment;
- equipment mounted in a remote avionics bay;
- control of equipment brought to the cockpit.

11.3.4 Military Aviation

Military aviation can be described as a hybrid of the previous three types of aircraft, plus an expansion on this. The expansion is basically on the emphasis on bigger, better, faster,

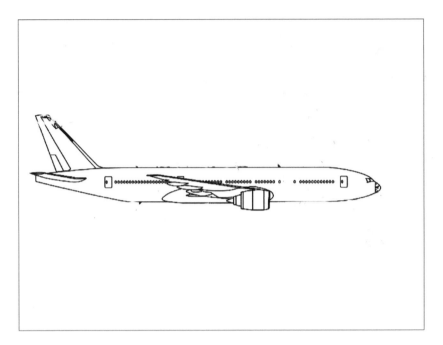

Figure 11.5   Commercial aviation aircraft.

more functional aircraft and on potentially an unlimited budget for national defence type applications. This brings a very unique functional specification that has to be taken on a case-by-case basis, but sometimes means incredible reliability, quality and weight/size environment requirements. It is amazing what a 10 dB in budget expenditure can provide, usually much more than 10 dB in technological performance. From an avionic perspective, usually the military is where all initiatives start from a leading edge technology perspective, and these aircraft types are even more likely in the future to drive the future communication systems and concepts (Figure 11.6).

Figure 11.7 gives a high level comparision between different aircraft groups and typical equipage.

11.4   Simple Avionics for Private Aviation

For a private aircraft, as discussed above, the communications avionics becomes very simple. In its most basic form a VHF communication transceiver is mounted in the cockpit, with associated dial and push-to-talk-operated microphone. Sometimes this VHF communication avionics is partially duplicated or duplicated to co-pilot space. The unit is usually supplied at 28 V DC, and a coaxial cable feeds RF output/input to a blade or even whip antenna (Figure 11.8). Note course, in some private pilot licence type applications such as gliders and microlights operating in limited airspace even this is not always required.

The corresponding navigation aids are outside the scope of this book but would generally include VOR direction finding equipment or maybe basic GPS, sometimes not even this. DMEs and SSR/ACAS transponders are not usually mandatory carriage in such light aircraft.

Figure 11.6 Military aviation aircraft. (Courtesy of John Mettrop)

11.5 The Distributed Avionics Concept

As was discussed previously, for the commercial applications, business jet applications and military there is a requirement to move to a distributed avionic topology, with control functions residing in the cockpit and bulk processing power and power electronics residing in a purpose built avionics bay with a secure data bus linking the two (Figure 11.9).

Probably the first place to start with distributed avionics is the data bus standards and then the power standards.

11.5.1 Data Bus Standards

11.5.1.1 ARINC 429 Standard

This legacy goes back to the 1970s with the 'data comms revolution' and networking computers together. The ability to communicate between different modules through a data bus that can be revised with mass scale integration MSI (i.e. through the invention of the transistor and integrated circuit), the communication avionics and associated modules became small enough and reliable enough, and it was for the first time economic and reliable to move the avionic function from the cockpit to a specialized rack to be known as the avionics bay or 'bays'. This had the advantage of freeing up valuable 'real estate' in the cockpit together with reducing cockpit heating effects.

	Private Aviation	General Aviation Business Jets	Commercial Aviation	Military Aviation
Example aircraft	Piper Warrior	Cessna Citation XLS Learjet 60XR	Boeing 737 Airbus 320	F-16 Hawker Harrier
Typical application	Private leisure	Business executive	Airline operation	Combat bomber surveillance
Key Avionic component drivers	Minimalist budget Certification	Certification Reliability Minimal weight and size Integrated test equipment Modularly changeable	Certification Reliability Minimal weight and size Integrated test equipment Modularly changeable Operational cost	Minimal weight and size Reliability Modularly changeable Stealth Technical Advantage
Typical communications equipage	1 × VHF Comm	2 × VHF Comm 1 × Satellite Comm 1 × HF Comm	3 × VHF Comm (Inc. Datalink) 2 × Satellite Comm 2 × HF Comm (Inc. Datalink)	2 × VHF Comm Military H F Comm 2 × JTIDS/MIDS Military Satellite Comm

Figure 11.7   Comparison of various aircraft avionic requirements.

The first equipment was interconnected to a 'data bus' using proprietary standards and formats (sometimes described in ARINC 419- 'Digital Data System Compendium' as emergent technologies). Soon the first mature standard emerged, generically known as ARINC 429, which was published as a first standard in 1977. This is still in regular operation today.

ARINC 429 provides an air transport industry standard for the transfer of aperiodic (or asynchronous) digital data between avionics system elements (or modules). The use of a standard allows interchangeability and flexibility of modules on and off the common communications data bus provided, if necessary, by different manufacturers (Figure 11.10). Of this standard,

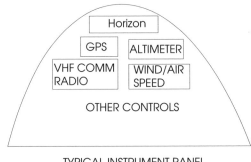

Figure 11.8   VHF mobile communications configuration for private aircraft.

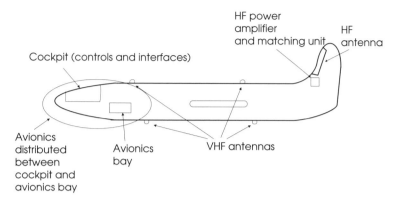

Figure 11.9 Distributed control, avionics.

- part 1 addresses the physical parameters (e.g. wiring, voltage, codes);
- part 2 addresses the format of words and encoding;
- part 3 deals with the data transfer protocols.

11.5.1.1.1 Physical System

The standard uses a digital information transfer system over single, twisted and shielded pair conductors. It uses single flow from output port of the device that wants to send information to all other elements likely to use information. (There is separation of input and output ports.)

11.5.1.1.2 Data 'Words'

The encoded word uses two's complement fractional binary or binary-coded decimal notation, with numbers coded as the ISO alphabet number 5.

Each word is a standard 32 bits with the last bit for parity (Figure 11.11).

There are typically a few hundred avionics 'elements' on a typical aircraft; each one has a unique code. (These unique codes are described in the standard.)

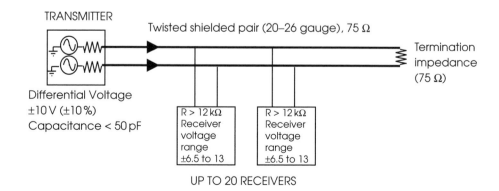

Figure 11.10 Example ARINC 429 data bus with avionic modules or 'elements'.

32	31	30	29	28	27	26	25	24	23	22	21	20	19	18	17	16	15	14	13	12	11	10	9	8	7	6	5	4	3	2	1
P	SSM	MSB											Data					LS B		SDI					label						

Figure 11.11   Standard word format for ARINC 429.

11.5.1.1.3 Modulation

The modulated signal uses bipolar return to zero signalling (this is tri state), with transmitting ±10 V (±0.5 V) as the states and 0 as the null changeover. When detecting these signalling levels, a receiver should indicate a positive state when voltage is between ±6.5 and ±13 V, and a non-state or null between these. (This gives considerable margin for distortions, etc.)

The Mark 33 protocol offers a number of different data speeds, from 'low' at 12–14.5 kbps to 'high' with speeds of 100 kbps.

Each type of communication has a predefined speed and rules that govern how much time (min and max) should be allowed between transmissions.

11.5.1.1.4 Transmission Line Loading

Any receiver should exhibit a differential resistance of 12 kΩ and a resistance to ground of >12 kΩ to the transmission line and a capacitance of 50 pF.

This enables modules to be added and taken from the bus seamlessly; i.e. the change in coupling resistance from one module does not affect the resistance to the other modules. (This is true for at least up to 20 devices)

The system is also designed so that it can continue to perform and transmitters should not be damaged in the presence of line faults and voltage transients, the detail of this is not discussed here. (Figure 11.12).

11.5.1.1.5 Limitations of the ARINC 429 Standard

In today's terms 10 kbps is considered slow. By comparison, the domestic home computer system with its LAN network (which traditionally defines the cheaper end of the equipment

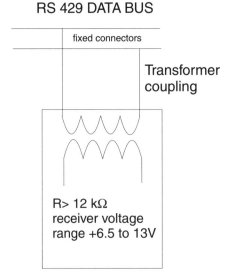

Figure 11.12   Load coupling.

market) speeds of 10 Mbytes/s and even 100 Mbytes/s are the most recent standards. This is 800 times faster than the maximum data speed achievable on ARINC 429.

Although experience has been good with ARINC 429 and it is a very proven and reliable standard, there is always room for improvement, and experience has shown its reliability weak spot is the potential for a single point of failure is the bus controller. Another lesser problem area is the ARINC 429 coupling which uses hard-wired transformer coupling. When faults occur it is usually evident from the loss of avionics function or from the Built-In Test Equipment (BITE). To put things in perspective, airline feedback on this standard has shown that in comparison to the card and avionics module level failures the bus problems are miniscule.

Finally, some concerns were made over the radio frequency interference (RFI) issues where precise carrier rates could cause interference into the LORAN C system, a potential way around this would be a move to fibre cabling with its intrinsic isolation and EMC qualities. This said, initial moves to fibre-based cabling structures were set back by maintenance considerations (duplicating test equipment for twisted pair and fibre systems) and there was even a migration back to the twisted pair standards, ARINC 429 and ARINC 629.

In addition LORANC is now all but obsolete.

11.5.1.2 ARINC 629 Standard

This standard became available in the mid 1990s. It was initially planned that this would ultimately supersede the ARINC 429 standard. It provides for typically 20 times faster bus data rates, and it uses multitransmitter data buses and also aims to get better reliability in that it removes the common point of failure of a single bus controller element. In addition, this standard can be used for twisted pair (in current or voltage mode) or fibre optic cables.

The hierarchy of this system uses subsystems that are coupled onto a global data bus via a terminal located in each subsystem. The intention is a single Tx/Rx terminal can accomplish this for each subsystem and hence minimize the number of connections.

Another 'improvement' over the previous 429 generation is through the non-intrusive, inductive coupling (in place of the transformer coupling with fixed connections) (Figure 11.13). This way no conductors need to be broken through the interconnection wire loom. (Inherently, a potential hot spot for faults is always in connectors.)

The data rate achievable is typically 2 Mbps over a twister pair wire bus or over a fibre optic bus. The speed was limited by using semiconductor technology from the 1990s. However, it is

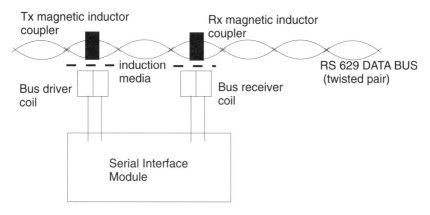

Figure 11.13 Coupling to ARINC 629 data bus.

now possible that data rates faster than this can be achieved (as a function of processor speeds only as the protocol is not itself speed limiting). The system is based on the Open System Interconnection Standard (OSI) reference model (IEEE standard 802.2). The bus medium can be current mode, voltage mode (over copper twisted pair cable) or fibre optic mode.

11.5.1.2.1 Protocol

It uses a carrier sense multiple access protocol as described in the Theory section of the book (and the same as VDL2) using bidirectional information flow on a line (much like a LAN network) and therefore no bus controller is required. This supports periodic and aperiodic data alike. The Basic Protocol and Combined Mode Protocol are intelligent protocols that mitigate problems with bus overload and excessive collisions by stepping into a slower rate aperiodic mode. The two cannot be used together on the same subsystem.

Basic Protocol. The bus can be operated in periodic or aperiodic modes. In periodic mode the transmissions are initiated by the bus controller. In aperiodic mode they are dependent on pre-arranged random unique 'terminal gaps'. Message lengths in aperiodic mode can be varied.

Combined Mode Protocol. In this mode there are three prioritized levels of bus access:

- level 1 – periodic transmissions of constant length;
- level 2 – aperiodic messages of short length;
- level 3 – periodic transmissions of max length 257 words.

11.5.1.2.2 Error Detection and Bus Integrity

Each terminal monitors its own and other transmissions with its built-in diagnostic error testing facility. Parity and cyclic redundancy checking are used at various system levels to protect data integrity. Physically, the bus specification requires that if any coupling stub is opened or closed it should not cause a bit error on the bus.

11.5.1.2.3 Experience of ARINC 629

As per feedback from airlines, reliability-wise ARINC 629 seems an improvement on ARINC 429, with very little reported problems directly associated with the bus wiring. In fact, at time of writing it was not possible to find any reports of problems with this protocol.

11.5.1.3   ARINC 659

This is the latest Boeing hardwire data loom standard, it is used on the latest Boeing 777 aircraft. It is an improved version of ARINC 629 with a higher speed of 60 Mbps. Many airlines are yet to report any faults with this system in its 10 years of operation.

11.5.1.4   Fibre-distributed Data Interface (FDDI)

Possibly inevitable was a move from copper twisted pair to fibre. The latest Boeing 777 implementations use this. It operates a dual 100-Mbps fibre LAN system with multiple copper 10-Mbps Ethernet sub-LANs (Figure 11.14).

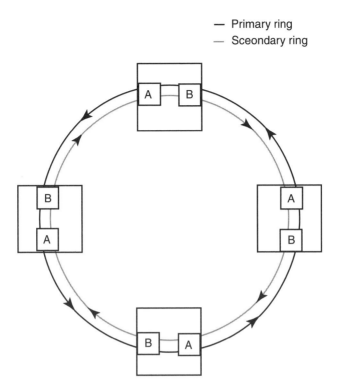

Figure 11.14 FDDI system.

There has been talk of decommissioning this system as it is too maintenance intensive with fibre and copper components, requiring an extensive amount of additional test equipment. In addition, the technology is becoming less available and consequently more expensive. There is a drive to replace the fibre components with an all copper 10 Mbps Ethernet version.

It should also be pointed out that this system is inherently very reliable, with an 'almost error free record' (There has been one reported instance with a broken fibre, but no failure of the electronics. In this instance of course the avionic function was uncompromised due to duplication.), and of course unspoken is its robustness to RFI aspects where the glass sectors are totally immune.

11.5.1.4.1 Airbus Developments

Airbus have announced its own version of the FDDI interface to be deployed on the new 380 aircraft. This will be a full-duplex, fully duplicated 100-Mbps main LAN with 10-Mbps, full-duplex, switched Ethernet communications subsystems.

11.5.2 *Power Supply System*

Traditionally the other common point, or common point of failure, in most communication systems is the power supply. This can be from primary generation through to rectification, or at the switching and fusing points or even on the circuit board.

11.5.2.1 Power Subsystem on an Aircraft

On most commercial aircraft it is fair to say that the AC and DC electrical buses are at least duplicated for reliability reasons. Even in single-engined aircraft it is common to see dual ignition systems and this practice is carried through to the avionics. In fact on the larger commercial aircraft quite often it is normal to triplicate generation sources (or even more) and the associated power system all the way to the avionics rack. At the avionics rack level, essential systems are either dual fed from this or more often in addition separate duplicated systems reside with one bus or the other.

11.5.2.2 Example The Boeing 777

Probably a good example to describe is a state of the art and yet mature system deployed on the Boeing 777 aircraft as standard configuration.

11.5.2.2.1 Primary AC (Figure 11.15)

The primary electrical system contains two integrated drive generators, one on each of the engines. Each of these has a 120-kVA direct to turbine generator. This generates the industry AC standard three-phase, four-wire, 400-Hz constant frequency 115 V/200 V (neutral connected to airframe). The super lightweight technology of these generation sets is an important feature, with the metric of interest being kVA delivered per kilogram of generator weight.

The system deploys a third emergency generator of 25 kVA attached to turbine gearbox to run 'essential services' only. On the ground the generation capacity can be temporarily supplemented by ground power AC supply.

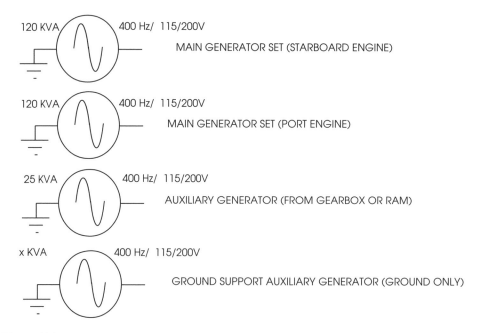

Figure 11.15 Primary AC generation on the Boeing 777.

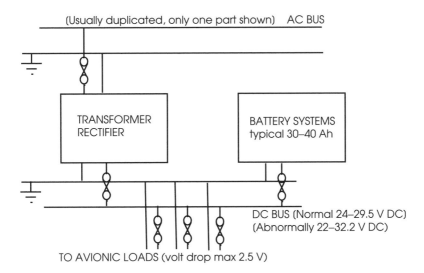

Figure 11.16 Boeing 777 DC distribution system.

It should also be pointed out that capacity-wise an aircraft should be capable of flying with one generator down and/or any part of this subsequent power system.

11.5.2.3 28 V DC

Typically there are two separate duplicated DC buses (i.e. four in total). The AC from the primary AC generation is fed through four separate (for reliability) rectifiers to supply battery backup for essential systems. Emergency RAM air turbine generators are also installed to generate electricity in the event of complete engine failure for the 'flight critical loads' (i.e. those systems required for the continuance of safe flying and landing). The negative pole of DC is also connected to airframe (earth) (Figure 11.16).

11.5.2.4 Flight Management System Monitoring of Circuit Breakers

Usually all instruments and avionics are fed by 28 V system with special circuit breakers for each used with status monitoring on the flight management system; i.e. if breakers trip, this is reported immediately to the pilots through the flight management system. The flight management system controls the automated load shedding function in the event of power failures.

ARINC 609 and ARINC 413A define the detailed requirement of typical electrical systems from good practice, capacity, over-voltage and fault perspectives.

Also RTCA DO 160E discusses the electrical 'environment' the equipment has to contend and cope with whilst still retaining performance, reliability and in the limit survivability. It has a detailed section on all the testing required to check this in Section 17. In Section 18 it also defines the audio systems on-board and aircraft and their susceptibility and limits to harmonic frequencies from the power system.

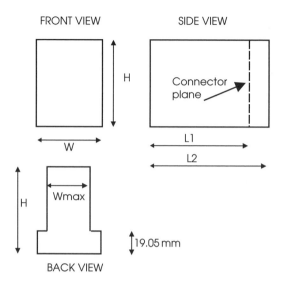

Figure 11.17 Standard ATR rack.

11.6 Avionic Racking Arrangements

The basic system for avionics racking was devised and is called the ATR units system. The origins of ATR are unclear as two parties are laying claim to the acronym. ATR can stand for 'Austin Trumbull Radio' or 'Aeronautical Transportation Racking' system (Figure 11.17).

The standard ATR rack has

- width 10.12 in. or 257 mm;
- length (long version) 19.62 in. or 495 mm;
- length (short version) 12.62 in. or 318 mm;
- height 7.62 in. or 193.5 mm.

A large number of variations to the basic rack are available including dwarf racks 1/4, $^1/_2$, 3/8, 3/4 width versions of the above. This is described fully in ARINC 404A and ARINC 600 specifications.

11.6.1 ATR and MCU

Another nomenclature for racking is a modular concept unit rack or MCU. Its relationship with ATRs is as described in Table 11.3.

The emphasis is on interchangeability of the avionic modules of the airlines (sometimes called line replaceable units or LRUs), i.e. the ability to swap out one unit from one manufacturer and slot in a new unit from a different manufacturer with the functionality of the avionics box essentially the same, and thus the avionic system is unchanged or minimal software configurations are required to swap them out.

Cabinets, racks and shelves to which the LRUs plug into are usually optimized in relation to the available real estate inside the aircraft equipment bay or cockpit; thus they are usually integral to the airframe manufacture and will usually be specified at the time of manufacture. (An adaptation later can be cumbersome as it will involve recertification.)

Table 11.3 ATR versus MCU standard rack sizes.

ATR rack size (ARINC 404A)	Equivalence in MCUs	L × L × W (in.)	L × L × W (mm)
1 1/2 ATR	12 MCU	12.76 × 7.64 × 15.31	324 × 94.06 × 388.88
1 3/8 ATR	11 MCU	12.76 × 7.64 × 14.01	324 × 94.06 × 355.86
1 1/4 ATR	10 MCU	12.76 × 7.64 × 12.71	324 × 94.06 × 322.84
1 1/8 ATR	9 MCU	12.76 × 7.64 × 11.41	324 × 94.06 × 289.82
1 ATR	8 MCU	12.76 × 7.64 × 10.11	324 × 94.06 × 256.8
7/8 ATR	7 MCU	12.76 × 7.64 × 8.81	324 × 94.06 × 223.78
3/4 ATR	6 MCU	12.76 × 7.64 × 7.52	324 × 94.06 × 191.02
5/8 ATR	5 MCU	12.76 × 7.64 × 6.21	324 × 94.06 × 157.74
1/2 ATR	4 MCU	12.76 × 7.64 × 4.90	324 × 94.06 × 124.46
3/8 ATR	3 MCU	12.76 × 7.64 × 3.58	324 × 94.06 × 90.94
1/4 ATR	2 MCU	12.76 × 7.64 × 2.27	324 × 94.06 × 57.66
1/8 ATR	1 MCU	12.76 × 7.64 × 1.01	324 × 94.06 × 25.65

Note that the MCU concept is taken from the 'short' ATR specification.

11.6.2   Cooling

The number of cooling holes in the avionics case will define the 'resistance' to airflow. So the cooling required can be carefully matched to what is available (Table 11.4).

11.6.3   Back Plane Wiring

The back plane interface of an avionics module and the corresponding rack seating are the functions of avionic function, avionic enclosure type, the connectors required on the rear of the avionics and the back plane wiring standard of the aircraft involved (the data on wiring and power supply wiring have already been described) (Figure 11.18). For most regular applications, the standard connector to be used is a MIL standard MIL-C-81659 connector. This

Table 11.4 Maximum weight and cooling specification.

Type	Maximum weight (kg)	Maximum thermal dissipation (W)	
		With forced cooling	No cooling
12 MCU	2.5	300	35
11 MCU	5.0	275	32
10 MCU	7.5	250	30
9 MCU	10.0	225	27
8 MCU	12.5	200	25
7 MCU	15.0	175	22
6 MCU	17.5	150	20
5 MCU	20.0	125	17
4 MCU	20.0	100	15
3 MCU	20.0	75	12
2 MCU	20.0	50	10
1 MCU	20.0	25	7

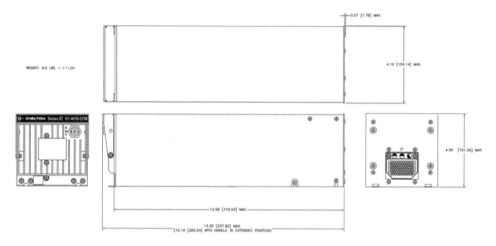

Figure 11.18 Back plane interface.

is a matrix connector of three and sometimes four or five columns (designated A, B, C, D and E) by 15 rows (designated 1–15); also there are connectors for coaxial cables and special signals.

11.6.3.1 Index Pin Code

In order to avoid the wrong avionics box being located in the wrong rack position, index pins are used at the rear that are unique to the avionics being used at a rack position.

11.6.4 Other Standards

Other lesser known standards have been emerging of late, particularly 19-in. variants to the ATR system, or VME and AIMS (airplane integrated management system) proprietary standards. These are found sometimes to offer a more flexible approach to the discipline of avionics and are attempts to modernize the racking arrangements with emphasis on large-scale integration and ever reducing weight and size.

11.7 Avionic Boxes

11.7.1 VHF Transceivers

Salient Features

- ARINC 716-11 defines a typical transceiver architecture and functionality;
- avionics unit located in avionics bay, controlled by the radio management system;
- more emphasis on maintainability, interchangeability than on size and weight;
- system parameters previously defined in ICAO Annex 10;
- transceiver should comply with ARINC 600 specification for 3 MCU form factor;
- size-1 shell ARINC 600 service connector;

Figure 11.19 Typical VHF transceiver module.

- service connections in middle plug, automatic test equipment connects to top plug and coaxial and power connections in its bottom plug. Index pin code 04 should be used (Figure 11.19).
- 27.5 V working;
- connects with standard NAV/COMM control panel (mounted in cockpit);
- power switch not in unit, continuous power to unit with circuit breaker contained on circuit breaker panel;
- grounded to ship's structure;
- complies with RTCA DO 160E – environmental conditions and test procedures for airborne electronics/electrical equipment and instruments.
- cooling, standard airflow rate at 13.6 kg/h of 40 °C air, dissipation not more than an average 62 W of power, air cool and resistance should not be more than 8-mm-pressure drop at 1.01325 bar
- frequency selection standard (described in ARINC 720-1).

11.7.1.1 Transmitter Specification

- Transmitter frequency offset is the ability of the transmit channel to be a few megahertz above the frequency the channel receiver is tuned to. Double-channel operation can be enabled by grounding the 'frequency offset enable' pin, either from control panel or another source. (Whether this offset feature is a future provision is still to be decided, but would require very infrequent change of the separation frequency – between Tx and Rx channels.)

- Power output under test conditions into a nominally 50 Ω resistive load should be 25–40 W on all operating frequencies and should be power rated to operate continuously (with forced air cooling previously discussed).
- Frequency stability generally better than 0.002 % for most steady-state conditions (oscillator stability DC voltage between 22 and 29 V, temperature –50 to +71 °C, humidity 10–95 %, pressure varied from sea level to 40 000 ft equivalent).
- Transmitter spurious should be less than –46 dBW (and in north America, less than –75 dBW for in-band spurious and –105 dBW for 108–118 MHZ band).
- Frequency response should be flat (within 6 dB) from 300 to 2500 Hz baseband input (this is for 25 and 8.33 kHz working).

11.7.1.2 Receiver Specification

- *Transmit to receive recovery.* From making or releasing the push-to-talk button, this should be <50 ms (to recover output power with squelch set to 3 μV, with Rx input level 10 μV modulated by 1000 Hz tone at modulation depth 30 %).
- *Sensitivity.* Voice circuit SNR > 6 dB with a 2 μV (hard) signal, amplitude modulated 30 % at 1 kHz.
- *Selectivity.* The noise pass-band filter passes signals for carrier ±8 kHz (for the 25-kHz system), with maximum attenuation 6 dB, >60 dB of attenuation should be achieved when modulated carrier departs >17 kHz from departed carrier (Figure 11.20). (This combined criteria facilitates offset carrier CLIMAX operation.).

Spurious out-of-band components and image frequencies should be rejected by at least 80 dB (and at least 100 dB, preferably 120 dB if they fall within the communications band). Cross-modulation products at least 10 dB down with respect to audio output.

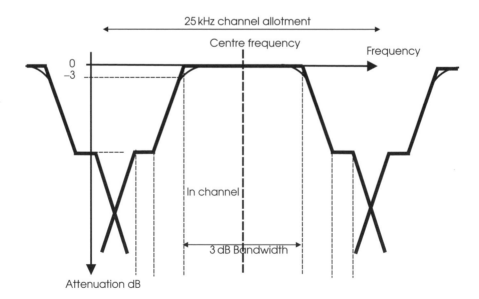

Figure 11.20   Receiver pass-band and rejection characteristics (selectivity curve).

Audio output, from 5 to 40 mW adjustable into a 600-Ω load.

Receiver gain – with 2 µV of signal, receiver produces at least 40 MW of power into 600-Ω load.

Automatic gain control – the receiver amplitude modulated output should not vary more than 3 dB with signals with dynamic input between 5 and 100 000 µV (and not more than 6 dB between 5 and 500 000 µV). Also receiver is set to be desensitized to pulse type interference or frequency modulation (FM) broadcast.

FM broadcast intermodulation interference – no degradation of performance should occur where third-order intermodulation products in the band of 118–137 MHz result from two or more FM broadcasting signals of –5 dBm or less mixing with the receiver.

FM broadcast desensitization interference – no degradation in performance should occur if the aggregate level of one or more FM broadcast signals across the VHF communications transceiver input terminals is less than –5 dBm.

Receiver frequency response should be from 312 to 1200 Hz. Post-detection response should be within ±6 dB from 300.0 to 6.6 kHz.

Distortion with 1000 µV signal modulated at 30 %, 1 kHz should be less than 5 %.

11.7.1.3 Navigation Communication Control Panel

As part of this the radio frequency control subpanel will usually have the following features (Figure 11.21).

The frequency selector will cover 118–137 MHz, with preset channels; for the 25-kHz working, this will give two decimal place display; for the 8.33-kHz working, it gives a three decimal place display.

The display is normally in 1/4 in. numerals or bigger. Under dual channel operation (i.e. when Tx and Rx are on different frequencies), the Rx channel is only shown.

There may or may not be a volume control for depending on airline/customer requirements.

No on/off control is provided.

Integral panel lighting is required for night-time viewing.

Sometimes (with custom design but not mandatory) it is required to be able to pass the control functions from one receiver to another. This is not covered in detail here and it is recommended that this option be looked at in manufacturers' specifications.

11.7.1.3.1 Example System

The VCS 40 VHF communication system includes an all-digital VHF transceiver built and designed to ARINC 700 series air transport specifications. The VCS 40 features a 20-W solid-state transmitter and can work with 25 kHz or 8.33 kHz channel spacing. The transceiver is tuned via the individual Series III control heads or ARINC 429 data bus, allowing it to interface to flight management systems and other standard 429 protocol systems to include the Chelton RMS-555 Radio Management system. The VC-401C transceivers and CD-402C controls are form, fit and functionally interchangeable with VC-401B and CD-402B units. The control head allows the operator to select 25 kHz or 8.33 kHz tuning steps using a knob position selector on the front of the control head.

Figure 11.21   NAV/COMM control panel.

- Channelization: 8.33 and/or 25 kHz Channel spacing; Mode: continuous transmit capability (at reduced power); Applications: built-in SELCAL and ACARS; Capability: optional stand-alone; Diagnostics: system self-test and diagnostics; Data bus: ARINC 429 digital data bus; Frequency of operation: 118.00–136.975 MHz; Standard frequency range: 118.00–151.975 MHz extended; Approvals: C-37C and C-38C (VC-401B and CD-402B), C-37D and C-38D (VC-401C and CD-402C).
- *Specifications:* Power requirements, 27.5 V DC; aircraft power, ±20 %; current requirements: receive, 0.5 A; transmit, 4.5 A, weight, 6.0 lb (2.7 kg); temperature, −67 to 158 °F (−55 to +70 °C); altitude, 55 000 ft (16 764 m); frequency range, 118.000–136.975 MHz or 118.000–151.975 MHz; channel spacing, 8.33 and/or 25 kHz; physical dimensions, 4.1 × 4.0 × 13.33 in. (104.14 × 101.60 × 338.58 mm); transmit power, 20 W.

11.7.2 HF Radios

The applicable standard for HF data equipment is ARINC 719-5.

11.7.2.1 Technical Specification

- For an SSB HF (upper sideband working) radio system, 3 kHz channelization in 1 kHz steps;
- Frequency selection by digital data ports (see ARINC 720);
- 2.8–24-MHz working;
- 1 kHz channelization tuning ability;
- Transmit to receive recovery typically better than 250 ms;
- Mode of operation – simplex operation;
- Receiver sensitivity – with 1-μV (hard) signal, the SNR > 10 dB and with a 4-μV(hard) signal 1 kHz tone 30 % modulated, SNR > 10 dB;
- Receiver selectivity – bandwidth at 6 dB down on centre > 5.5 kHz and bandwidth at 60 dB down on centre < 12 kHz;
- Frequency accuracy – Tx and Rx frequency within 20 Hz;
- Tx spurious – >60 dB down on centre (preferably >80 dB in band 2.8–24 MHz);
- Voice output – variable, 5–40 mW into 600-Ω load;
- Receiver gain – with 2-μV signal modulated 30 % at 1 kHz will produce > 40-mW output into 600-Ω receiver;
- Receiver frequency response – <6 dB over frequency range 300–2500 Hz, sharp cut-off above and below is desirable;
- AGC – receiver operating power change <6 dB, with input signals from 5 to 100 000 μV;
- Advisable when SELCAL mode is used Rx sensitivity is set to max (squelch circuit disabled);
- Signal linearity – with two tone test between threshold and 20 000 μV; and the intermodulation products should be at least 40 dB and preferably 50 dB below;
- A separate SELCAL data output with source impedance of less than and independent of voice circuit and associated squelch and processing should be provided 300 Ω;
- Receiver frequency response should be designed so that no more than 3 dB difference occurs between SELCAL tones between 300 and 1500 Hz.

11.7.2.2 Transmitter

- Tx power output >200 W (optional 400 W for valve devices) peak envelope power (PEP) into 52 Ω;
- Tx power output >100-W carrier power;
- Transmitter rated for continuous (50 % on) duty cycle for 5 minutes;
- Audio input: This is per standard ARINC 538A and 559A specifications, with a load resistance of 150 Ω and filtering network of 200 Ω, band limiting the signal.

Tx output spectrum shaping: (Figure 11.22)

350–2500 Hz (flat gain with max 6-dB variance in pass-band);
>30 dB attenuation at 100 Hz below carrier and 2900 Hz above carrier;

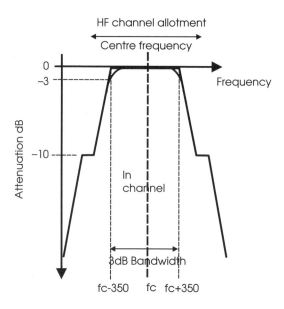

Figure 11.22   Transmitter spectrum shaping.

>40 dB at 3.1 kHz below carrier and >5.9 kHz above carrier;
Components with frequency lower than 6.1 kHz and higher than 8.9 kHz should be attenuated
   by at least 54 dB and preferably 60 dB or more (with the exception of harmonics of carrier);
Any emission from harmonic should be down at least 43 dB below PEP;
All intermodulation distortion and spurious radiation should be at least 60 dB below PEP.
Transmitter distortion at PEP with sinusoidal modulating inputs should not exceed 10 %;
Tuning interlock required if two HF transceivers are installed on one aircraft to prevent inad-
   vertent co-frequency selection and problems.

11.7.2.3   HF Physical Specification

Uses a 6 MCU form factor;
Size-2 shell ARINC 600 connector;
With top connector for testing (on lab bench only),
middle connector for main service connections
and bottom connector for power and coaxial connections,
index pin code 10 should be used to define the back plane and the correct matings (Figures 11.23
   and 11.24).

11.7.2.4   Power

The HF transceiver is a high-power device and hence it is designed to take its 3-phase power
from the 115 V AC bus. The avionics module should be designed to dissipate no more than
260 W of power with a forced cooling of 57.2 kg/h at 40 °C air applied. (For the state-of-the-
art high-power 400 PEP transmitter this can be increased to 540-W dissipation and 120 kg/h
forced air cooling.) (Figure 11.25).

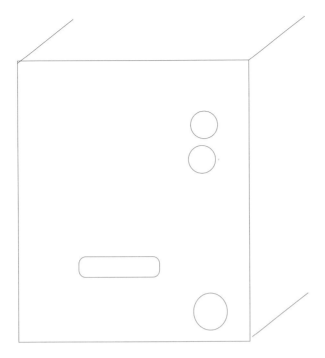

Figure 11.23 Back plane connectors for HF radio module.

11.7.2.5 HF Built-in Test Equipment

As with most avionics, this is required to detect a minimum of 95 % of faults and failures; i.e. minimum of red (= fault) and green (= OK) lights on front display of avionics.

11.7.2.6 HF Antenna Tuner and Coupler

The function of the antenna tuner is to tune the antenna system (as it varies in impedance with frequency across the whole band 2.8–24.0 MHz) to the coaxial transmission line at a nominal 52 Ω and to keep the voltage standing wave ratio (VSWR) < 1.3:1.0.

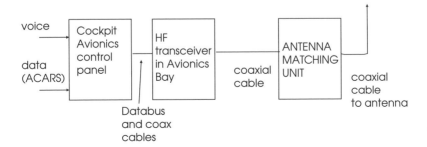

Figure 11.24 Wiring schematic for HF radio system.

Figure 11.25   HF radio configuration.

Automatic control of SSB R/T power output through sensing the VSWR is required to keep the spurious level down. Full PEP can only be achieved when the antenna and transmission line system are optimally matched. Power handling of tuner >650 W.

The tuner/coupler is usually located in remote parts of the aircraft where it can be as close as possible to the antenna. It is preferred if this is rack mounted with easy maintenance access; however, this is not always possible. Usually the antenna tuner unit is individually pressurized and there are normally contact closures inside the unit that close under correct pressure conditions. Transmitters should be inhibited from transmitting unless these contacts are closed.

11.7.2.7   Dual System Interlocks

Where one antenna is being used for two transceivers, interlocks are required to prevent isolation and front-end damage to either receiver.

11.7.2.8   HF Data Radio

HF data radio is defined in ARINC 719 and ARINC 635 specifications.

Data in. In 'data in hot' and 'data in cold' data facility into HF avionics modules, input impedance is preferably 600 Ω (can be 100 Ω). No wave-form processing should be applied. (This is applicable to the voice circuits only.) Datalink audio frequency response should be 350–2500 Hz (flat response ±6 dB).

Data out. Open circuit output of 0.5 V RMS, with source impedance of 100 Ω, with a receiver input either AME or SSB; for SELCAL tones, should not be more than 6 dB difference between any two tones.

11.7.2.8.1   Examples

Rhode and Schwarz XK516

- the Airborne Voice/Data Radio is designed for use in commercial aircraft. The system provides conventional voice and high-speed air-to-ground, ground-to-air, and air-to-air data communication over long distances;
- used on Airbus and Boeing, McDonnell Douglas;
- voice per ARINC 719 specification, data per ARINC 635 specification;
- 1800 bps data;
- 2× ARINC 429 data buses;

- the Collins HFS-900D HF Data Radio provides the air transport industry with a low-cost, long-range datalink system for fleets operating in oceanic, polar and remote land areas. When paired with the Collins CPL-920D Digital Antenna Coupler, the HFS-900D high-frequency datalink (HFDL) system provides the high-reliability system needed for HFDL applications. In addition, Collins HFDL can be paired with a single satellite communications (SATCOM) system to provide maximum datalink communication availability. Because these two systems have different propagation characteristics and link failure modes, HFDL and SATCOM systems together provide a 99.8 % availability, meeting FAA's air traffic management requirement.

Rhode and Schwarz R&S M3AR

- the transceivers of the R&S M3AR family are software-re-programmable and are among the smallest and most lightweight VHF/UHF airborne transceivers worldwide. It is possible to download several electronic program after modulation (EPM) waveforms and use them alternately. The architecture of the R&S MR6000L ensures form, fit and function replacement for existing AN/ARC-164 radio systems;
- multiple EPM waveform options:

  — SATURN;
  — HAVE QUICK I/II;
  — SECOS.

- software-re-programmable ;
- 30–400-MHz frequency coverage;
- voice and data;
- 10 W AM to 15 W FM;
- 8.33 kHz and 25 kHz channel spacing;
- weight <4.0 kg.

11.7.3 Satellite Receiver System Avionics

A typical block diagram of a satellite receiver (Inmarsat M standard) is shown in Figure 11.26. It should be noted that the active electronics of both the satellite receiver and transmitter are integrated into the flat mount antenna. This is to maximize the link budget as discussed in both Theory and System sections of this book (Figure 11.27).

11.7.3.1 Receiver Specification

According to ARINC 741, for the low-gain receiver a figure of merit of better than –26 dB/K should be achieved and for the switched beam high-gain amplifier a figure of merit of –13 dB/K or better should be achieved.

11.7.3.2 Size Specification

HGPA type 1 (with heat dissipation of 125 W) should be 4 MCUs;
HPGA type 2 (with heat dissipation of 250 W) should be 8 MCUs.

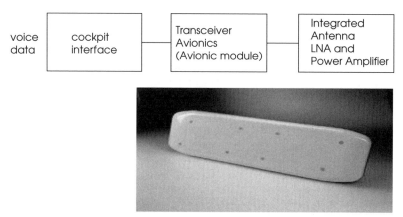

HIGH GAIN FUSELAGE TOP MOUNT ANTENNA SYSTEM

Figure 11.26   Satellite system.

11.7.4   Other Equipment

It is impossible to cover all the combinations and permutations of avionics and as such just the generic equipment such as VHF, HF and satellite equipment have been covered in detail above. The military equivalents become very involved and also very specialized and specific and also secret, so they have not been included in the scope of this book; similarly pre-certification avionics such as telemetry equipment are also outside the scope of this book.

Finally, in passing, it is worth looking at the total avionic connectivity of a typical aircraft and this is shown below for the 'generic commercial airliner' (Figures 11.28 and 11.29). Avionics is a specialized discipline in its own right and we have only just scratched the surface or given a general introduction.

11.8   Antennas

11.8.1   VHF Antennas

These can be (a) whip antennas on light aircraft and (b) blade antennas, which are more common on the faster and higher performance jets.

Figure 11.27   Avionic boxes.

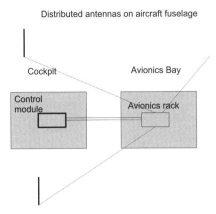

Figure 11.28 Schematic layout of generic commercial airliner avionics.

11.8.1.1 Whip Antennas

Whip antennas are appropriate only for light aircraft up to just over 100 knot. Physically they are end-fed, 1/4-wave antennas. The characteristic impedance is usually 50 Ω; however, this and the performance can change slightly with strain (with speed). VSWR can vary up to 3:1, typically rated to 30 W EIRP (effective isotropic radiated power). Polarization is vertical with these antennas are respect to the antenna. (Although consideration has to be given to pitch and roll of the aircraft, in practice this changes negligibly.)

The general radiation pattern is omnidirectional in the horizontal plane, this can practically be slightly modified with the adjacent airframe metallic elements. Severe attenuation nulls apply when going through significant parts of airframe or wing.

Figure 11.29 Communications management unit.

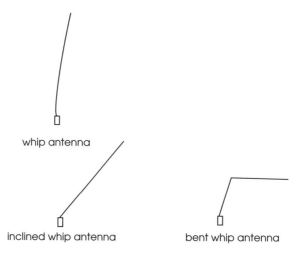

Figure 11.30 Whip antenna, inclined whip or bent whip antennas.

It is also possible to have variations of the basic whip antenna: an inclined whip antenna or a bent pole antenna, which are usually designed for twin prop aircraft or medium size aircraft with higher ratings to typically 250 knot and above. The electrical characteristics of a bent pole antenna usually provide a wider frequency band of operation.

There are many different variations manufactured. Usually they are supplied with BNC connectors that are embedded in the mount matings for direct connection to associated coaxial cables. Usually mounting is done to penetrations to the fuselage of 2, 3, 4, 5 or 6 holes with a suitable weather/pressure sealing gasket arrangement (Figures 11.30 and 11.31).

Advantages:

- can be relatively cheap;
- easy to install and swap out, with no specialist tools required.

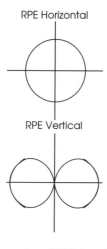

Figure 11.31 Typical radiation pattern envelope (RPE) for whip antennas.

Figure 11.32   Typical blade antennas.

Disadvantages:

- cannot be used at high jet speeds, where conductors overbend;
- can be easily damaged by hail, birds and other contact.

11.8.1.2   Blade Antennas

In the limit, a bent antenna turns into a blade antenna and these are usually encased in polyurethane enamel. A blade antenna can take many shapes and forms. Electrically, it is a 1/4-wave, end-fed antenna with an impedance of $50\,\Omega$, a VSWR typically up to 2.5:1.0, rated to about 50-W EIRP. For VHF communications, vertical polarization is required (Figure 11.32).

The general radiation pattern is omnidirectional in the horizontal, sometimes slightly modified with the adjacent airframe metallic elements. Severe attenuation nulls apply when going through significant parts of airframe or wing.

Typically these antennas are rated well above 500 knot. Many different variations of antennas are manufactured. The leading edge is strengthened to protect from damage from birds, hail or other hazards. Usually they are supplied with BNC connectors that are embedded in the mount matings. Mounting is done to penetrations to the fuselage of 2, 3, 4, 5 or 6 holes, with a suitable weather/pressure sealing gasket arrangement.

Advantages:

- robust and rated for high-wind speeds;
- no appreciable change to antenna pattern with speed.

Disadvantage:

- can be relatively expensive.

Blade antennas for VHF communications are usually mounted on the top of the fuselage and on the underneath of the fuselage to give the best all-round coverage with manoeuvring.

11.8.1.3 Compound Antennas

It is possible to integrate a number of antennas into the physical casing on one antenna. This is to reduce the co-site issues or to minimize antenna external antenna cluttering (Figure 11.33).

11.8.2 HF Antennas

11.8.2.1 Wireline

An important property of the HF antenna is that it needs to operate at wavelengths between 10 and 100 m (see the Theory section).

A number of different topologies are deployed. Traditionally and on smaller aircraft the typical wire antenna is used as an end-fed (from half-way down the fuselage); the other end is insulated from the tail or in small aircraft grounded. Lightning protection is incorporated.

For the higher performance jets, the antenna can produce drag and the reading angle relative to the fuselage should be minimized.

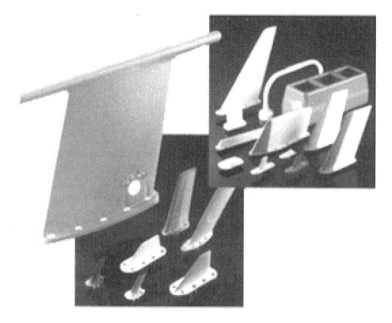

Figure 11.33   Example compound antenna.

Advantages:

- low-cost solution;
- easy to retrofit.

Disadvantages:

- wind drag;
- parameters can vary considerably over the band, affecting antenna efficiency and reliability;
- risk if wire breaks (need zero-tension release mechanism).

11.8.2.2 Probe Antennas

A probe antenna is a horizontal conductor, approximately 2–3-m long, placed on the far-forward edge of a wing or tail and is electrically isolated from the wing. It exhibits electrical properties close to those of a quarter-wavelength antenna. Again this is a susceptible point for lightning strikes and an arrester needs to be put at the wing connector to protect the antenna feeder and coupler.

Advantages:

- reasonably economical to manufacture and install;
- flexible in that you can have more than one on an aircraft;
- good reliability and performance;
- reasonable impedance characteristics across the band of interest (Figure 11.34).

Disadvantages:

- usually the end of a wing tip or tail experiences most vibration and temperature fluctuations;
- antenna impedance and pattern can be influenced by proximity to airframe.

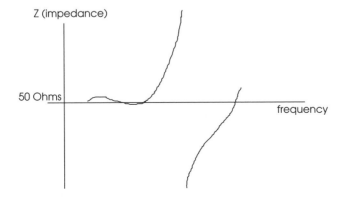

Figure 11.34 Typical antenna impedance characteristics with frequency.

11.8.2.3 Cap Antennas

This is where the antenna is an integral 'flush' part of a tail or wing section. It has similar electrical properties to the probe antenna. Again it is prone to lightning, although arguably less because it does not have a sharp end. The electrical characteristics are less than that of a quarter-wavelength antenna.

Advantages:

- no additional wind drag;
- good reliability;
- good radiation pattern.

Disadvantage:

- inherently more expensive as it involves structural changes to aircraft (unless captured in design and airframe manufacture).

11.8.2.4 Shunt Antennas

A shunt antenna is similar to a notch or cap antenna except that it is installed using insulating materials in place of normally metal sections. For example, on the leading or trailing edge of the wing a long slot is cut. Its electrical properties are similar to those of a 1/4-wave, end-fed antenna. (Its physical length should be less than 1/2 a wavelength.)
   Advantages and *disadvantages* are similar to those of cap antennas.

11.8.2.5 Notch Antenna

Notch antennas are usually, as their name implies, 'notch shapes' cut, for example, deep into the airframe at the lower leading edge of the tail or stabilizer.
   Advantages and *disadvantages* are similar to those of cap and notch antennas.

11.8.2.6 Antenna Couplers

Antenna couplers connect the antenna to the antenna-tuning unit. They are directional and protected with relays and dummy loads in case of fault of antenna or transmission lines causing fluctuations in VSWR.

11.8.3 Satellite Antennas

There are two critical constraints regarding satellite antenna systems on-board an aircraft.

1. For a passive receive antenna, the received signal power is a function of antenna diameter.
2. Pre-amplification at an antenna before a transmission line gives a better overall system noise figure and hence a better performance. This factor is critical with the satellite communications receiver system on-board an aircraft in addition to the severe antenna size limitation.

Physically is not possible to have a 3.7-m diameter antenna on the aircraft; however, it is possible to use a 'steerable' active antenna. To achieve the highest figure of merit of the antenna system, it will require a pre-amplifier immediately after it. This gives two good reasons that the satellite antenna system is an active device needing its own power supply (steerable antenna and then pre amp).

11.9   Mastering the Co-site Environment

From Figures 11.35–11.37 it can be easily seen how quickly the aircraft becomes cluttered with antennas for the different applications it has to operate, in particular with larger aircraft.

It has been estimated that there can be typically over 20 radio systems operated on one conventional jet aircraft on the many different aviation frequencies. This can have upwards of 30 or 40 antennas. By their very nature these antennas are restricted in proximity to each other by the physical length, width and height of the aircraft, and it is further restricted by the environment and aircraft operations (i.e. high power and active HF; and hence satellite antennas need to be away from fuel systems).

Also the location of antennas is usually such that they can be fit for purpose; for example, satellite antennas should be placed on the upper side of the aircraft so that under normal conditions they are facing up and can carry out their function; conversely, landing system receiver antennas, the altimeter antennas, and terminal Aeronatical communication antennas are placed underneath. For VHF communications antennas should be placed on top and underneath to give the optimal coverage and for the best line of sight for the operation being undertaken (i.e. underneath for approach, on top for ground communications). This in itself places some very significant compatibility issues on the systems. For illustration, consider the GNSS receiver systems which usually work way down below the noise floor in the L band (around 1400 MHz). These are extremely prone to interference from DME in close proximity or from 10th harmonics from a VHF system or inter-modulation products from other of the adjacent transmitters. Consequently their antenna location needs very considerable thought. The same is true of all the antennas on-board an aircraft: Each time one is moved, it affects co-site compatibility with all the other systems. So an on going interactive optimization is required.

Even more problematic can be the systems that operate in the same band. An example of this might be VDL 4 or VDL 2 and the voice communication systems, and to date it has been found that even with careful planning and location, interference can be caused from one

Figure 11.35   Typical small private aircraft antennas (VHF and HF on top of fuselage).

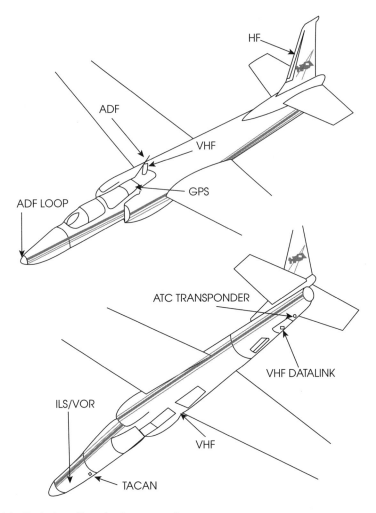

Figure 11.36 Typical medium size jet antenna layout.

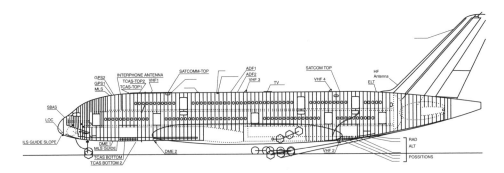

Figure 11.37 Typical large jet antenna layout.

system into the other. It also has some serious implications for future communication systems when potentially there will be co-band operations going on in VHF, L band and 5 GHz band, possibly exacerbating the problem; and of course, the ever important space of real estate on-board an aircraft is always running out. Good luck to future system planners as this aspect gets even more complex!

Finally, it is worth pointing out that as a 'space-saving' measure, it is possible to use combined antennas, i.e. where a number of systems are coupled onto one antenna.

11.10   Data Cables, Power Cables, Special Cables, Coaxial Cables

Interconnecting all the avionics system components that are in the cockpit, in the avionics bay or in the antenna systems mounted externally, are cable systems. These can be data systems with twisted pair specification as described (e.g. ARINC 429 or ARINC 629 standard) or fibre optic buses.

Then there are coaxial cables required usually to provide low-loss signals from the avionics power amplifiers to the transmitting aerial (or the converse return path); there are specialist cables to connect, say, headphones and voice-activated microphones in the cockpit to the avionics in question; and finally there are various grade power cables depending on application and current draw.

The important characteristics of all these cables are that they meet the environmental requirements previously specified, they are low-loss for those applications involved and they provide the most flexible configuration or future configuration potential. Also the combined weight of all the cables on board an aircraft is very significant and in future wireless solutions using RLAN type interfaces are being investigated (For non-safety critical functions such as entertainment system connection, this is very conceivable for the future).

The detail of the cables is embedded in the various specifications and equipment manufacturers' technical literature and is not discussed further here.

11.11   Certification and Maintaining Airworthiness

11.11.1   Certification

Certification is the collective term for the approvals required for an aircraft to become 'airworthy.' This is essentially a national process of what an aircraft has to have complied with in its manufacturing process together with its final fit out and testing, before it can be considered airworthy. Emphasis is on the airframe, engines, all component parts and subsystems for 'safe operation'.

Once airworthy, the aircraft has to continue with various airworthy compliances mandated by the National Civil Aviation Authority (or equivalent regulating body) and where appropriate the JAAs (usually for flying internationally around Europe and into North America). Any modification (be it from changing component parts, drilling minute holes in racks, sub-frames or the fuselage up to major retrofits or changing engines) has to be re-assessed as to what degree of re-certification is required. There is a process involved that defines this in each state and in the JAA. The avionics are a sub-component and a vital subsystem element to this whole process.

11.11.2 EUROCAE

EUROCAE was formed in 1967 by the European Civil Aviation Conference (ECAC) and was tasked with the preparation of minimum performance specifications for airborne electronic equipment. In time the European National Airworthiness Authorities were recommended to take EUROCAE specifications as the basis of their national regulations. Today, the JAAs consider compliance with EUROCAE documents as means to satisfy Joint Technical Standard Orders and other regulatory documents.

11.11.3 Master Minimum Equipment List

This is a national register usually held by the aviation regulating authority for the country in question or by the JAA and it defines the minimum amount of equipment or functional equipment that is required by an aircraft before it is considered airworthy and can take off. It also defines the maximum amount of equipment that is allowed to be out of service and leaving aircraft still serviceable.

From this a clear emphasis can be seen on reliability, interoperability and maintainability of avionic equipment.

Further Reading

1. ARINC 404A – Air Transport Equipment Case and Racking
2. ARINC 404B – Connectors, Rack and Panel, Rectangular Rear, Release Crimp Contacts
3. ARINC 413A – Guidance for Aircraft Electrical Power Utilisation and Transient Protection
4. ARINC 421 – Guidance for Standard Subdivision of ATA Spec 100, Numbering System for Avionics
5. ARINC 429 – Mark 33 Digital Information Transfer System Part 1, Functional Description, Electrical Interface, Label Assignment and Word Formats
6. ARINC 542A – Digital Flight Data Recorder
7. ARINC 566A-9 – Mark 3 VHF Transceiver
8. ARINC 594A – Mark 2 airborne SELCAL System
9. ARINC 606A – Guidance for Electrostatic Utilization and Protection
10. ARINC 609A – Design Guidance for Aircraft Electrical Power Systems
11. ARINC 619-1 – ACARS Protocols for Avionic End Systems
12. ARINC 620-4 – Datalink Ground System Standard and Interface Specification
13. ARINC 622-4 – Datalink Application over ACARS Air Ground Network
14. ARINC 600-15 – Air Transport Avionic Equipment Interfaces
15. ARINC 629 – Multi Transmitter Data Bus (Part 1 Technical Description)
16. ARINC 631-1 – VHF Digital Link Implementation Provisions
17. ARINC 632-2 – Gate-Aircraft Terminal Environment (Gatelink Ground Side)
18. ARINC 634 – HF Datalink system guidance design material
19. ARINC 635-4 – HF Datalink Protocols
20. ARINC 664 – Aircraft Data Network Part 1 – Concepts
21. ARINC 701-1 – Flight Control System
22. ARINC 702-6 – Flight Management Computer System
23. ARINC 702A-1 – Advanced Flight Management Computer system
24. ARINC 714-6 – Mark 3 Airborne SELCAL System
25. ARINC 716-11 – Airborne VHF Transceiver
26. ARINC 719-5 – Airborne SSB HF System
27. ARINC 720-1 – Digital Frequency Function, Selection for Airborne Electronic Equipment

28. ARINC 724-9 – Communication Addressing and Reporting System (ACARS)
29. ARINC 727-1 – Airborne Microwave Landing System
30. ARINC 741 – Aeronautical Satellite Communication System
31. ARINC 750 Part 4 – VHF Data Radio
32. ARINC 801 – Fiber Optic Connecters
33. ARINC 802 – Fiber Optic Cable
34. ARINC 803 – Fiber Optic System Design Guidelines
35. EUROCAE ED 14E – Environmental Conditions and Test Procedures for Airborne Equipment
36. EUROCAE ED 23B – Minimum Operational Performance Specification for Airborne VHF Receiver – Transmitter Operating in the Range 117.975–137 MHz
37. EUROCAE ED80 – Design Assurance Guidelines for Airborne Electronic Hardware
38. EUROCAE ED90A – Radio Frequency Susceptibility Test Procedures
39. EUROCAE ED107 – Guide to Certification of Aircraft in a High Intensity Radiated Field (HIRF) Environment
40. EUROCAE ED108 – Interim MOPS for VDL Mode 4 Aircraft transceiver for ADS–B
41. RTCA DO 160E – Environmental Conditions and Test Procedures for Airborne Equipment
42. www.chelton.com
43. www.rsd.de
44. www.rockwellcollins.com
45. www.inmarsat.com
46. www.airbus.com
47. www.boeing.com

All ARINC standards are published by www.arinc.com
All EUROCAE standards can be found at www.eurocae.eu

12 Interference, Electromagnetic Compatibility, Spectrum Management and Frequency Management

Summary

This chapter looks at the related topics *interference* and *electromagnetic compatibility* (EMC) and the disciplines that stem from these, namely *spectrum management* and then last in the chain *frequency management*.

It starts with definitions of interference, how a radio engineer can protect against this through careful planning and defining EMC criteria between systems. It then looks at Spectrum management and, in particular, at the ITU international radio regulations and how these form the basis of international 'radio law'. Internationally, a cooperation is set up by each National Radio Administration signing the World Radio Conference (WRC) treaties with the intent that these agreed regulations are mandated and enforced into National Law where ultimately the legal responsibilities lie.

The radio regulations (RRs) form a dynamic framework for accommodating new radio systems whilst at the same time protecting existing interests. Finally, frequency management

Aeronautical Radio Communication Systems and Networks   D. Stacey
© 2008 John Wiley & Sons, Ltd

is about the day-to-day running of the allocations through allotment and assignment of specific radio equipment to a service.

(As a prerequisite to this chapter a background understanding of the three As, described in Chapter 2, together with a general understanding of the system level is assumed.)

12.1   Introduction

Till now it has always been assumed that systems operate in an ideal environment, with dedicated spectrum allocated only to the system under consideration. Unfortunately, this is not the case. The spectrum resource is a finite resource and with radio applications growing daily and arguably going through a number of revolutions of late (particularly mobile radio and mobile data applications); the spectrum is becoming 'filled up' or congested. This is particularly true of the spectrum below 5 GHz, which is in great demand due to its propagation characteristics, making it most suitable for many applications. Consequently, it is necessary to manage this process and best guard existing systems against this perceived congestion and growing interference. Ultimately, this is the function of the national radio regulators, and collectively, this is the ITU-R international radio regulations. (The *Red Book* of Radio Regulations is available at www.itu.org.)

The RRs are an agreed compromise between all world radio administrations as to how the radio spectrum is cut up and managed, how existing systems are best protected, how old systems can be phased out and spectrum handed back, and how new systems can be best introduced. This spectrum management process is an ongoing dynamic one, theoretically done on an equitable, rational and economic basis, and importantly one where change is relatively slow (For example, clearing a band of old or legacy functions and reorganizing it for a new application or service can take decades: For some services it can happen much faster.)

12.2   Interference

So, what is interference? From the ITU-R radio regulations[1] interference is defined as 'The effect of unwanted energy due to one or a combination of emissions, radiations, or inductions upon reception in a radiocommunication system, manifested by any performance degradation, misinterpretation, or loss of information which could be extracted in the absence of such unwanted energy'.

It can be considered as a 'sister' to noise: both limit the performance of a radio system. Usually noise will always be there and is totally random and uncontrollable (it is sometimes called *white noise* in that its power spectrum is uniformly spread). On the other hand, interference is usually controllable in some way and it does not necessarily have a flat power spectrum, although this is often assumed in the first case for ease of analysis.

Sometimes the two can be confused or even lumped together, maybe because interference under certain conditions can be said to behave like noise and it can initially be analysed from a mathematical equivalence.

12.2.1   Sources of Interference

These can be generally split into two parts (Figure 12.1): accidental or inadvertent interference and intended or purposeful interference.

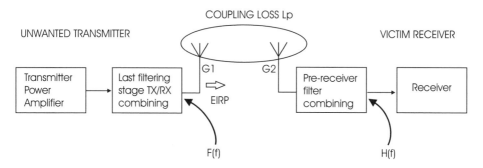

Figure 12.1 Interferer – victim scenario.

12.2.1.1 Accidental or Inadvertent Interference

This type of interference can come from a number of different sources. Some of them are as follows.

Electrical disturbance. This is when electrical power equipment, such as transformers, induction heaters and industrial power plants, for example, under steady-state operating conditions or under fault transients produce harmonics that are in the pass-band of a wanted receiver signal or where poor isolation between electrical power and communications equipment exist. This had been a big problem in the past, but generally in the developed nations this foreseen eventuality can be protected against with the use of various EMC regulations.

EMC issues with adjacent equipment. This is similar to the electrical disturbance but instead involves electronic and communications equipment being operated close to a sensitive receiver (Good examples of this would be a handheld GSM phone operating close to a GPS receiver or ultra-wide band (UWB) devices in close proximity to radar services.) or the interactions between all the antennas on an aircraft with numerous and complex avionics.

Cable leaks and radiation. Under some conditions a poorly installed or antiquated cable TV network, in particular, poorly made or corroded connections in such a network, can become an aerial, which launches RF cable power usually on the carrier of the TV network. This has been found to cause frequent problems in the VHF and UHF bands, usually by an aggregation of leaks.

Intermodulation products. This is when two or more fundamental frequencies are in an environment where coupling can be achieved, and due to non-linearities in the coupling, multiples of various fundamental frequencies as well as their additions and subtractions can be made (see Chapter 2 for detailed mathematical combinations). Usually with some analysis and precautions, this kind of interference can be identified and filtered out.

Self-interference. This is when a component of the user's transceiver system causes problems into the wanted signal. Usually this kind of problem is ironed out at manufacture, but it can be a site-specific problem caused by faulty equipment.

Intrasystem interference. This is an extended case of the self-interference where one transceiver system may interfere with an identical transceiver system elsewhere. Normally this is a frequency coordination issue which falls under the frequency management processes.

Intersystem interference. This is where one transceiver system can interfere with another transceiver system operating in a different mode. It may be co-channel or adjacent channel but could be further away than this. Usually under steady-state conditions this problem is managed by frequency coordination; if this is not appropriate, then EMC or spectrum management processes are utilized.

Licensing mistake or coordination problem. Not all that uncommon is when a licence has been erroneously issued or the users are found not to be operating their equipment within the terms of the licence.

Non-existent, poor or out-of-date EMC protection criteria. As the take-up of services is always changing, sometimes the regulatory provisions do not properly reflect the requirements needed for proper EMC to exist. EMC regulations need continual update and monitoring as services and environments change. This last aspect is sometimes a hard issue to identify and certainly involves a longer time frame fix than most of the predecessors. In the EMC specification, various tools can be used to properly protect the victim receivers. Some of the tools used are specifying masks of interferer, required signal-to-noise ratio criteria and protection limits for receivers, C/I ratio, co-channel limits, adjacent channel limits, adjacent band limits, polarity requirements, spurious transmitter masks, duty cycles, statistical probabilities.

Multiple effects or unexplained sources. Sometimes multiple effects of all of the above causes can be seen, which can get complicated to identify and resolve, and of course, sometimes the source is never properly identified.

12.2.1.2 Intended or Purposeful Interference

This type of interference is caused by purposeful human intervention. Following are its examples:

Jamming and malicious jamming. Jamming is seen when a known interference source is applied intentionally directly into the path of the victim receiver. Good examples of this are observed under war conditions where an enemy jams radars or the very sensitive GNSS/GPS signals. Or it is often possible to buy GSM phone jammers to alleviate the anti-social behaviour of loud mobile phone users, although usually operating such devices is illegal.

   Of more concern is the malicious intent by a few minor individuals to actually cause harm to radio users by purposefully trying to block transmissions for essential safety services. In the extreme this could be terrorism of sorts.

Pirate radio. Operating pirate radio is illegal but purposeful and sometimes because of poor engineering or even non-engineering and coordination such channels can be found over the top of wanted services or harmonics and intermodulations of such a system get into an adjacent service. This is a frequent and growing problem for VHF communications and navigation bands in a number of regions.

12.2.2 Interference Forms

Interference can take many forms – some easy to identify, some more sporadic and harder to troubleshoot. Some of them are as follows:

Continuous source. This is obviously the easiest to locate.

Non-continuous, sporadic. These are harder to locate and troubleshooters may have to wait for further bursts to examine them.

Non-continuous, once off. These are almost impossible to locate and most of these go down on fault-reporting sheets as unknown.

Non-white noise. This interference can cause interesting side effects such as intermodulation or harmonics getting in through the pass-band of the receiver filter even though the main frequency of interference is outside this band (e.g. UWB, broadcasting systems).

12.2.3 Immunity and Susceptibility

Immunity can be described as the degree to which a device is protected from interference (usually in technical terms, by the way of an immunity specification giving levels and parameters). Susceptibility is the inverse of this and the degree to which a receiver is liable to be interfered with.

Immunity and susceptibility are a function of

- nature of interference;
- signal-to-noise ratio required for the receiver;
- where the interference level sits in relation to the noise floor (sometimes this is given as an I/N value (e.g. for radars);
- the usual receive power into the receiver, the receiver threshold and the difference between these (the fade margin);
- sophistication of the receiver in rejecting interference and the active countermeasures.

Example 1

Figure 12.2 shows a spectrum mask for new UWB service that is planned to be introduced over the next few years (these limits are taken at 3 m from the source under a measured bandwidth of 1 MHz). (UWB can be assumed to radiate across the bandwidth of the receiver.)

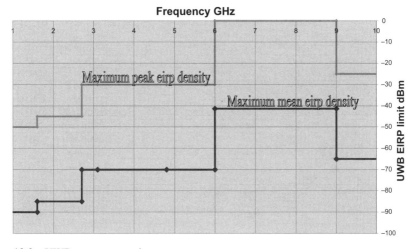

Figure 12.2 UWB spectrum mask.

Figure 12.3 shows normalized (to 1 MHz) receiver sensitivities at distances from the UWB device sketched on the same mask.

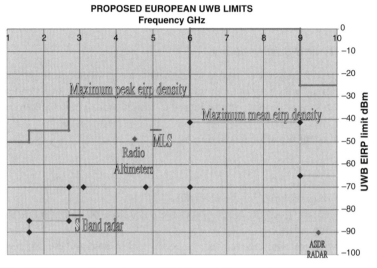

Figure 12.3   UWB mask with receiver sensitivities and distances shown.

Q1. Which services would be expected to be more prone to interference?

Q2. How would these figures be in reality?

Answer

Q1. Basically the radar services can be seen to be the most sensitive, with radar altimeters and MLS (microwave landing system) also potentially at risk. The degrees to which these are at risk can be measured by the number of decibels that fall under the line. This number is their susceptibility (see Figure 12.4).

System	Designed receiver threshold dBm/MHz	dBm equivalent at antenna/ bandwidth	Protection ratio (Safety of life factor + protection required)	Max interfering signal power dBm/bandwidth	Source of Reference
VHF 25 kHz voice (air)	30 µV/m	−94/25 kHz	20	−114/25 kHz	ICAO ANNEX 10
VHF 25 kHz voice (ground)	20 µV/m	−90/25 kHz	20	−110/25 kHz	ICAO ANNEX 10
VHF 8.33 kHz voice (air)	30 µV/m	−94/25 kHz	20	−114/25 kHz	ICAO ANNEX 10
VHF8.33 kHz (ground)	20 µV/m	−90/25 kHz	20	−110/25 kHz	ICAO ANNEX 10
VDL2,3 (air)	30 µV/m	−94/25 kHz	20	−114/25 kHz	ICAO ANNEX 10
VDL2,3 (ground)	20 µV/m	−90/25 kHz	20	−110/25 kHz	ICAO ANNEX 10
VDL4 (air)	75 µV/m	−82/25 kHz	20	−102/25 kHz	ICAO ANNEX 10
VDL4 (ground)	35 µV/m	−88/ 25 kHz	20	−108/ 25 kHz	ICAO ANNEX 10
UAT (air)	Not specified	−98/1 MHz	20	−118/1 MHz	ICAO ANNEX 10
UAT(ground)	Not specified	−98/1 MHz	20	−118/1 MHz	ICAO ANNEX 10
HF comms and data (air)		(−110)/3 kHz	15	(−130)/3 kHz	ITU RR A.27
HF comms and data (air)		(−110)/3 kHz	15	(−130)/3 kHz	ITU RR A.27
Satellite (air)		TBD		TBD	ITU RR
Satellite (ground)		TBD		TBD	ITU RR

(Example only, number still under validation)

Figure 12.4   Susceptibility of systems.

> Q2. This work needs some more careful study and further practical testing by the spectrum management process. Maybe looking at the protection mechanisms available to mitigate against this problem, this could use statistical techniques, automatic detect before transmitting mechanisms, automatic transmit power control; maybe consider in more detail antenna pattern rejection characteristics that have not been considered so far. Then there is always a proposal to amend the regulation to properly reflect the problem if one of the previous provisions cannot fix the compatibility problem.

12.2.4 Testing for Interference

There are a number of practical ways to test interference. Some of the following suggestions may apply:

- Turn off the transmitter end of the wanted system and see how the received power level changes. If this does not drop below the noise level, there is likely to be interference. Check this with a power meter and spectrum analyser (best to do when stationary, if possible).
- Direction-finding equipment may be needed to triangulate on interference point from two or more geographic locations. If the facility exists, use a directional antenna to make sweeps of problematic area with wanted transmitter turned off and look for peaks in the received signal. This can help locate direction of source. (This of course uses stationary interfering source.) (Figure 12.5).
- Look with a spectrum analyser at the received power spectral density.
- Isolate the equipment from the radio environment (dummy paths) and see if the same problems persist.

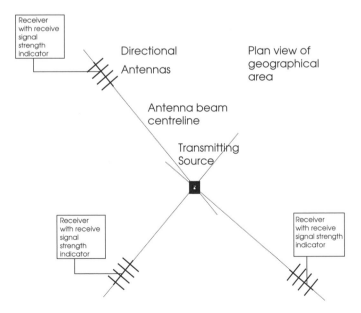

Figure 12.5 Direction finding.

- For fixed links switch the polarity.
- For AM signals listen to the channel with a demodulator and detect language and broadcaster heading
- Most European administrations have facilities to either send out monitoring equipment and direction-finding equipment for serious problems or they can even fly routes to measure unwanted spectrum components (e.g. for cable TV networks).

12.3 Electromagnetic Compatibility

This is the detailed study of the spectrum used for a receiver system and what limits are acceptable for that receiver. This section looks in detail at wanted and unwanted spectrum components.

12.3.1 Analysis

Defining the process of EMC between two (or more) devices can be broken down into the following stages of analysis. If the two (or more) devices effectively operate in the environment along side each other, they are said to be electromagnetically compatible. If either of the devices causes degradation to the performance of the other, they are said to be electromagnetically incompatible. The degree to which they are either compatible or incompatible is studied by the disciplines of spectrum management and frequency management.

Stage I – Define wanted system limits. The first place to start is by analysing the wanted transmitter–receiver path. At this stage, it is important to write down all the parameters

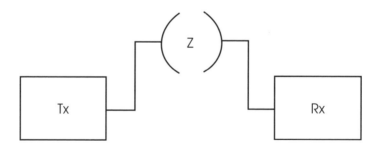

Figure 12.6   Wanted transceiver paths.

required for the radio link to function properly. These may include transmitter powers and mask and minimum receiver powers (sometimes called the threshold) and masks (Figure 12.6). A traditional link budget table can be set up.

In particular, a received spectrum mask will be defined, which in turn dictates the maximum limits of any adjacent emitters and thus a transmitter mask for co-channel, adjacent channel and adjacent band services (Figures 12.7 and 12.8).

Stage II – Characterize nature of interferer. In this next stage, it is important to define in the same terms the characteristics of the interfering radio system (or the system with which compatibility is required); these will usually be all the same criteria such as transmitted spectrum, received spectrum of the interferer's receiver system and the masks (Figure 12.9).

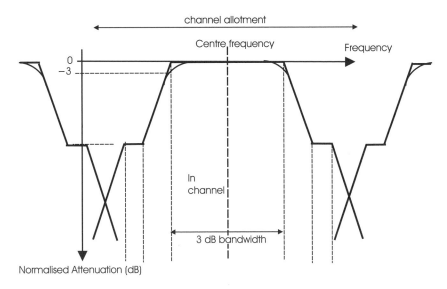

Figure 12.7 Example receive mask.

Stage III–Analyse relationship between wanted and unwanted interferers. This is to lay out the two systems in relationship to one another. In particular, spacial distances between each of the systems are important. The relative free space path losses, the directivity of the antennas both towards wanted and unwanted signals and the relationship of receiver bandwidths are also important. Normalization can be done assuming white noise equivalence as described in Chapter 2 (Figure 12.10).

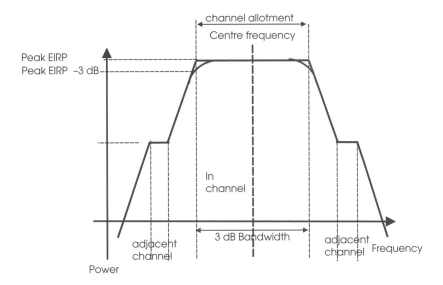

Figure 12.8 Example emission mask.

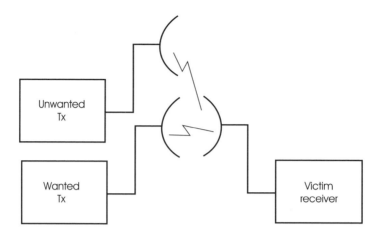

Figure 12.9 Wanted and unwanted transceiver paths.

Stage IV – Set the criteria. Knowing the minimal receiver powers required at receivers (or their antenna inputs) enables the engineer to define limits (usually but not always) below this for maximum incident interference levels into the device from the adjacent sources. The margin between these limits is sometimes called the protection margin or EMC margin.

12.3.2 Out of Channel, Out of Band, Spurious Emissions

In addition to the above discussion, transmitting sources usually have a bandwidth beyond which their spectral emissions are seriously attenuated. These are called the out-of-band emissions, which also have some defined limits. Way beyond the bandwidth of the transmitter, there can be significant spectral components. These anomalies are sometimes called spurious emissions and can be caused by, for example, harmonic components from some of the main frequencies. There are some limits for these spurious emissions as well (Figure 12.11).

To illustrate this a bit better, the defined spectrum masks for the VHF communications band (118–137 MHz) transmitters are shown.

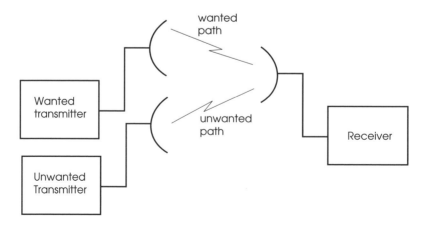

Figure 12.10 Example relationship between wanted and unwanted interferers.

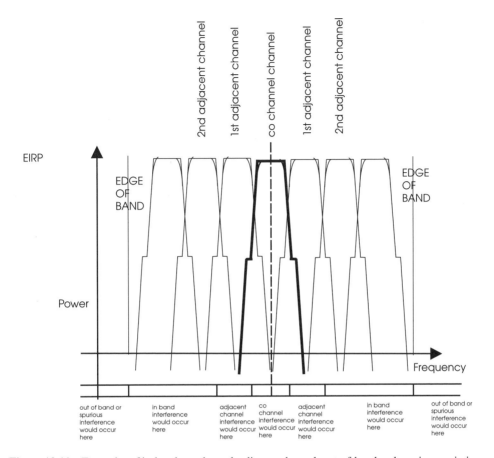

Figure 12.11 Examples of in-band, co-channel, adjacent channel, out-of-band and spurious emissions.

12.3.3 EMC Criteria

There is probably no one place where EMC criteria are defined. In addition, these can have moving goalposts in versions of regulations, code of practice, and standards, which are always being updated. Some of the obvious references for EMC regulation are ITU Radio regulations (usually the high-level requirements only) and the ICAO SARPs (for specific aviation systems) in Annex 10. Some of the detailed requirements are given in National Documents (for the United Kingdom, the MPTs or British Standards), or by various international regulatory bodies such as EUROCAE, RTCA, ETSI, CEPT ECC (Figure 12.12).

A matrix of these parameters and how they relate can be constructed as part of the EMC analysis, it should be noted that this matrix will be dynamic and ever changing as systems are modified and regulation is updated, mainly because it is a subject of ongoing study and validation. Historically, this information is not always available. Also, it is forever the subject of modification and update as part of the spectrum management and frequency management processes. Attention should be given to the source of criteria as unfortunately conflicts can be found when comparing different EMC documents.

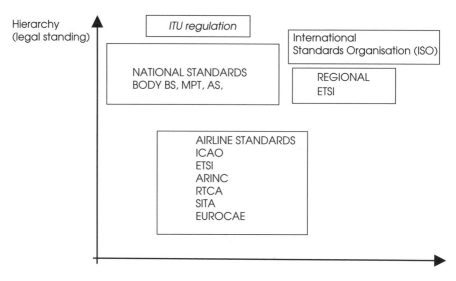

Figure 12.12   Example tree of EMC regulation.

12.3.3.1   Building a Compatibility Matrix

Part of the process of showing EMC is the building of a compatibility matrix. This is where the service or system under threat (sometimes called the victim) is considered in relation to the other incumbent or proposed new systems.

12.3.3.1.1   Example

A typical such matrix for a new system in the 5091–5150-MHz band may look as shown in Table 12.1.

12.4   Spectrum Management Process

Spectrum management is the tactical, political and long-term process of knowing how allocations are made to the ITU-R radio regulations. It is the dynamic world of ensuring adequate protection and EMC between a number of allocations.

Table 12.1   Example victim versus interferer matrix.

		Victim		
		MLS	AMS(R)S	FSS feeder links	Ground-based communication system
Interferer	MLS	⊠			
	AMS(R)S		⊠		
	FSS feeder links			⊠	
	Ground-based communication system				⊠

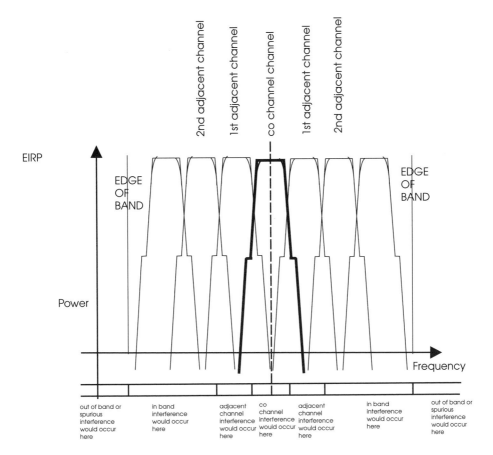

Figure 12.13 Differentiation between co-channel, adjacent channel and adjacent band compatibility.

12.4.1   Co-channel Sharing and Adjacent Channel and Adjacent Band Compatibility

Co-channel sharing is where systems are operating on the same frequency where this is appropriate. *Adjacent channel compatibility* (this is when one system is operating on a neighbouring adjacent channel in the same frequency allocation. *Adjacent band compatibility* is when two systems are adjacent to each other on different frequencies operating in different band allocations (Figure 12.13). In the extreme, 'adjacent' can even mean multiple allocations away.

12.4.2   Intrasystem and Intersystem Compatibility

Another dimension to this is the nature of the sharing systems (intrasystem, intersystem). To explain the concepts and terms, an example is offered up (Figure 12.14). This example is a real one and it looks at the ITU table of allocation between 118 and 137 MHz VHF to the aeronautical mobile (route) service, AM(R)S.

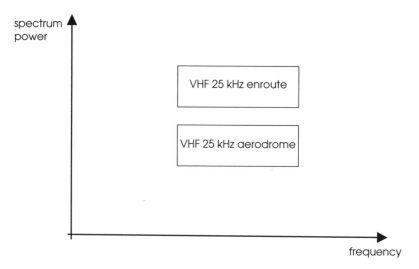

Figure 12.14 An example of intra system spectrum allocation.

12.4.3 Intrasystem Criteria

These are used when two same independent transceiver systems are operating in the same band allocation. An example could be two VHF 25-kHz AM(R)S systems. this is the most easy situation to coordinate and it is usually carried out by the frequency management process after initial coordination (i.e. the first time only) through the spectrum management process.

12.4.4 Intersystem Criteria

These are used when in one frequency band a number of different system types are operating; that is when the band is 'shared'. The status of services in a band has different priorities: Primary means one service has priority over another service; secondary means it is second in priority. (Co-primary can also exist when the compatibility rules are clearly defined and both systems are said to have equal status and priority).

12.4.4.1 Two Aviation Systems

One scenario is when two of the systems in one band are aviation systems. A good example of this would be the radionavigation band 108.000–117.975 MHz where VORs and VDL 4 datalinks have a provision. Under this scenario, sharing criteria between the two systems have been defined. This is usually done within ICAO as both systems are ICAO-standardized systems.

12.4.4.2 Two Systems: One of Them Not Aviation Safety of Life

A more complex scenario is when one of the two systems falls outside the control of aviation or jurisdiction of ICAO, for example the band 960–1215 MHz. In this band there are DME

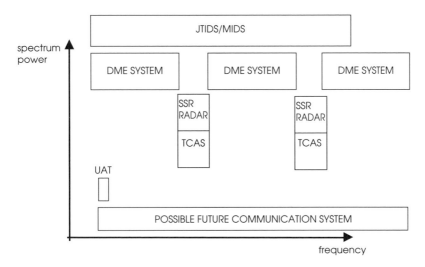

Figure 12.15 Intrasystem and intersystem EMC and their responsibilities.

devices that fall under the self-coordination criteria of ICAO. There are also proposals for a new UAT system at 977 MHz and a future AM(R)S system, as described in Chapter 9, which are both to be ICAO-standardized systems. If these were the only issues, they could have been handled internally by the ICAO. However, there are also national radars allocations in this band and also the non-ITU-standardized, non-disclosed JTIDS/MIDS that are not controlled by aviation or ICAO.

The intersystem coordination criteria for all of these systems are more complex and go well beyond the jurisdiction and interests of aviation in their effects and impacts. It becomes a broader subject of study of the radio administrations concerned and ultimately the ITU-R (in this case study, group 8B) to look at the spectrum management mechanism for this band (Figure 12.15).

Increasingly, as the spectrum becomes more congested, spectrum sharing is being encouraged and the luxury of exclusive use of band allocations by individual services is becoming more and more a thing of the past. This adds a degree of complexity to the whole spectrum management and EMC process.

12.4.5 WRC Process and the Review and Amend Cycles

A WRC is held usually every 4 years. This is the opportunity to amend the radio regulations and to change the high-level EMC criteria defining how every service is protected and should operate.

Feeding this process are national and regional preparations (regional preparatory groups include CEPT, APT, CITEL, the former CIS block group, the African Group and the Arab Group). Also feeding this are the studies conducted by the ITU study disciplines, which in turn are being fed by regional and national standardization bodies such as ETSI, CISPRA, EUROCAE, CEPT, ICAO. The exact relationship between all these bodies is specific to the radio system under discussion and is dynamic in nature as the EMC compromise is reached.

Ultimately of course the responsibility is at a national level, though for aviation issues ICAO tries to coordinate its spectrum management policy through ICAO AMCP-WG F. The amend and review process of all of these organizations varies but is usually a continual reiterative and cyclic process, leading up to the next WRC where things are ratified as much as possible.

12.5 Frequency Management Process

In contrast to the spectrum management process, frequency management is the final stage to the allocation, allotment and assignment process and mainly involves the last two elements. It involves applying coordination rules (previously defined as a shared effort between the spectrum management and frequency management processes) that define how frequency allocations can be made within one service (intrasystem protection) to ensure that interference (mainly self-interference) is minimized.

The allotment and assignment processes are usually done by local administrations, civil aviation authorities or in some cases ANSPs. For aviation channels, these are centrally coordinated by ICAO regional offices.

12.5.1 Example

Consider the frequency management issues surrounding the VHF band 117.975–137 MHz in Europe:

- The coordination criteria for this band are clearly defined in the European region by ICAO Air Navigation Plan COMM 2 Table (Table 12.2).[2] This document details agreement reached between states on all VHF assignments in 117.975–137-MHz bands. It is published yearly by the ICAO Paris Office.
- In its edition for 2007, there are over 11 000 entries in the ICAO database, of which over 1500 are for area control services. This number is a function of the geographical area, number of aircraft average and peak movements (sometimes described as the peak instantaneous air count) and the number of departures/arrivals in each sector and the number of en route transits.
- The ICAO database structure contains all the allocations in a country, by geographical order, by frequency, and of deleted assignments. It also details geographic information of sector boundaries.

12.5.2 Emergency Frequency (Three-channel Guard Band Either Side)

Channels assigned to upper airspace may also be assigned to lower airspace, not vice versa.

Co-Channel Criteria: High-Level Summary
- Circular service coordination distance is set to five times radius from boundary of circle or horizon whichever lessor;
- Non-circular – coordination distance is set to horizon;
- For valid assignment both required services and existing services must coordinate;

Table 12.2 AM(R)S VHF channel allotment of 117.975–137 MHz for Europe.

From	To	Channels (25 kHz)	Guard bands	Allotment
118	**121.4**	136		International and national mobile services
119.7	119.7	1		Reserved for regional guard supplementary tower and approach services
121.5	121.5	1	±3 channels	Emergency frequency
121.6	122	15		International and national aerodrome surface communications
122	123.05	42		National aeronautical services
122.1	122.1	1		Reserved for regional guard supplementary tower and approach services
122.5	122.5	1		Reserved for regional light aircraft
123.1	123.1	1	±1	International S&R (ATIS allowed in guard bands)
123.45	123.45	1		Air to air communications for remote and oceanic
123.5	123.5	1		Reserved for regional light aircraft
123.15	123.6917	21		National aeronautical mobile service
129.7	130.8917	47		National aeronautical mobile service
130.9	**136.875**	**239**		International and national aeronautical mobile service
131.4	131.975	23		Reserved for operational control communications
131.525	131.725	8		Reserved for ACARS datalink
136.8	136.875	3		Reserved for operational control communications, but no new assignment
136.9	136.975	3		VDL

In bold is the set.
Note also between 136.5–136.975 not for the use of 8,33.

- Circular service separation DOCs must be five times the range of larger DOC;
- Circular area services will be calculated using the radio Lorizon method etc. (for full information see ICAO COMM 2).

Adjacent channel criteria
- Circular to area > 0 NM;
- Area to area > 10 NM;
- ATC inside B service is OK, provided the distance between TXs = 10 NM.

Service	Description	Example
A	Area of service	ACC
B	Broadcast service (no airborne portion)	VOLMET, ATIS
C	Circular service	
E	European service (blocked from use)	EF 11.5, SAR 123.1
U	Unprotected service	AS, OPC

Designated operational coverage (DOC) range (NM) and height (FL).

There is an increasing inability to allocate new assignments; currently, a few hundred labelled 'P-provisional' = unresolved.

12.5.3 SAFIRE (Spectrum frequency information repositary)

A new database tool is being developed by Eurocontrol to facilitate on line, real time frequency coordinate between adjacent States.

Further Reading

1. ITU Radio Regulations, Articles Section VII, Article 1.166, 2003, Geneva WRC
2. ICAO Air Navigation Plan – European Region, Part X, Aeronautical Telecommunications (Com) Supplement 2 – COMM Table Com – 2. www.paris.icao.int
3. ICAO Standards and Recommended Practices, Vol 10
4. EUROCAE ED107 – Guide to Certification of Aircraft in a High Intensity Radiated Field (HIRF) Environment, March 2001, published by www.eurocae.eu
5. RTCA DO 160E – Environmental Conditions and Test Procedures for Airborne Equipment

Appendix 1

Summary of All Equations (Constants, Variables and Conversions)

Wave physics

$$v = f\lambda \tag{1}$$

Decibels

$$\text{Power dB (unit } x) = 10 \, \log_{10}(\text{unit } x) \tag{2}$$

$$\text{Amplitude dB (unit } x) = 20 \, \log_{10}(\text{unit } x) \tag{3}$$

$$E \, (\text{dB } \mu V/m) = 20 \, \log_{10} E(\mu V/m) \tag{3a}$$

Power flux density and field strength

$$\text{Power flux density (PFD) } S \text{ or } \Psi \, (W/m^2) = P_{Tx}/4\pi d^2 \tag{4a}$$

Effective isotropic antenna area

$$A_e = \frac{\lambda^2}{4\pi} \quad (m^2) \tag{4b}$$

Aeronautical Radio Communication Systems and Networks   D. Stacey
© 2008 John Wiley & Sons, Ltd

$$P_{Rx} = \frac{\lambda^2}{4\pi} \times PFD \quad (W) \tag{4c}$$

$$P_{Rx} = \frac{\lambda^2 P_{Tx}}{(4\pi d)^2} \quad (W) \tag{4d}$$

$$L_{fspl} = 32.44 + 20 \log_{10}(d_{km}) + 20 \log_{10}(f_{MHz}) \tag{4e}$$

$$\text{Power flux density } S = \frac{E^2}{Z_o} \quad (W/m^2) \tag{5}$$

$$Z_o = 120\,\Pi \text{ or } 377\,\Omega \text{ (the impedance of free space)} \tag{6}$$

$$S(\text{dB W/m}^2) = E(\text{dB }\mu V/m) - 145.76 \tag{7}$$

$$E(\text{dB }\mu V/m) = P_{Tx}(\text{dB W}) - 20 \log(d_{km}) + 74.8 \tag{8}$$

$$P_r = E(\text{dB }\mu V/m) - 20 \log(f_{MHz}) - 107.2 \tag{9}$$

$$P_r \text{ is isotropically received power (dB W)}$$

Earth geometry, horizon distance and *k* factor

Horizon distance

$$d_1 \text{ (km)} = (2Rh_1)^{0.5} \tag{10}$$

$$D \text{ (km)} = d_1 + d_2 = (2Rh_1)^{0.5} + (2Rh_2)^{0.5} \quad \text{(for two aircraft)} \tag{10a}$$

$$D \text{ (km)} = d_1 + d_2 = (2kRh_1)^{0.5} + (2kRh_2)^{0.5} \quad \text{(for two aircraft with } k \text{ factor)} \tag{11}$$

Distance between two points on Earth's surface

$$D = 2R \sin^{-1}[(\sin^2((\text{lat1} - \text{lat2})/2) + \cos(\text{lat1})\cos(\text{lat2})$$
$$\times \sin^2((\text{long1} - \text{long2})/2)]^{0.5} \quad \text{[note angles in radians]} \tag{12}$$

HF

MUF – Maximum Useable Frequency. This is a median value statistically. The chances at the time of validity of an HF channel frequency being 'open' above this value are 50%.

HPF – Highest Possible Frequency. This is the HF frequency above which statistically there is only a 10% chance of the frequency being 'open'.

FOT/OWF – Optimum Working Frequency. This is the highest HF frequency that gives an 85% likelihood of the channel being open.

FOE – Critical Frequency. Highest frequency that is returned in ionosphere. Calculated or found by sounding.

$$Lt = 32.45 + 20 \log f + 20 \log p\phi - Gt + Li + Lm + Lg + Lh + Lz \tag{13}$$

Doppler

Therefore

$$f_{\text{apparent}} = f_{\text{carrier}} \times (C + y)/C \quad \text{(Hz)} \tag{14}$$

where y is relative speed in m/s.

Modulation equations
DSB-AM

$$\text{fc}(t) = A \, \cos(2\pi \text{fc} t + \Phi) \tag{15a}$$

Let

$$A = K + \text{fm}(t) \tag{15b}$$

$$\text{fm}(t) = a \cos(2\pi \text{fm} t) \tag{15c}$$

So

$$\text{fc}(t) = (K + a \cos(2\pi \text{fm} t)) \cos(2\pi \text{fc} t + \Phi) \tag{15d}$$

$$m = a/K \tag{15e}$$

SSB-AM

$$\text{fc}(t) = 0.5m \, \cos((2\pi (\text{fc} - \text{fm})t)) - \text{upper sideband SSM} \tag{16a}$$

$$\text{fc}(t) = 0.5m \, \cos((2\pi (\text{fc} + \text{fm})t) - \text{lower sideband SSM} \tag{16b}$$

Digital modulatiion
ASK

$$1 = \text{turn tone on}, \ 0 = \text{turn tone off} \tag{17}$$

FSK

$$1 = \text{transmit frequency f1}, 0 = \text{transmit frequency f2} \tag{18}$$

BFSK

$$\text{Bandwidth} = 2B(1 + \beta) \tag{19}$$

$$B_{\text{s}} = 1/\text{t1} \tag{19a}$$

$$\beta = \Delta f/B \tag{19b}$$

No point in increasing the modulation index beyond 1, $\Delta f = B$.

Relationship between bit and baud

$$R\text{(bits/s)} = M \text{ (no of signalling states)} \times r \text{ (baud or signalling rate)} \tag{20}$$

Shannon's theory

Theoretical limit for error-free transmission

$$C = B \log_2(1 + S/N) \qquad \text{(bits/s)} \tag{21}$$

where
    C = channel capacity in bits/s
    B = bandwidth in Hz
    S/N is signal-to-noise ratio
    Always $R < C$; if $R > C$, errors will occur.

Erlang's theory

$$A = YS \tag{22}$$

where
    A = traffic in erlangs
    Y = mean call arrival rate (calls per unit time)
    S = mean call holding time (same units as y)

CDMA theory

$$I = (k - 1)C$$

$$C/I = 1/(k - 1) \quad \text{(dB)} \tag{23a}$$

Io is the interference power density in dBW/Hz, and W is the signal bandwidth in Hz.

$$\text{Io} = I/W \quad \text{(dB W/Hz)} \tag{23b}$$

Similarly if R is the data rate of the signal in bits/s,

$$Eb = C/R \quad \text{(J)} \tag{23c}$$

Linking all three of these equations together,

$$k - 1 = I/C = \text{Io}W/C = (W/R)/((C/R)/\text{Io})) = (W/R)/(Eb/\text{Io}) \tag{23d}$$

For large values of k,

$$k \approx (W/R)/(Eb/\text{Io}) \tag{23e}$$

$$\text{Coding gain} = 10 \log(\text{spreading factor}) \tag{23f}$$

$$\text{Spreading factor } W/R \text{ ranges from 4 to 256} \tag{23g}$$

Normalization to a bandwidth

$$P2(\text{dB W/B2 MHz}) = P1(\text{d BW/B1 MHz}) + 10\log B2 - 10\log B1 \qquad (24)$$

Antennas

$$\text{Isotropic} = 0 \text{ dBi} \qquad (25)$$

$$\text{Dipole} = 2.14 \text{ dBi} = 0 \text{ dBd} \qquad (26)$$

$$5/8\,\lambda \text{ vertical antenna, gain peaks close to 4 dBd} \qquad (27)$$

$$\text{Dish antenna } G\,(\text{dB}) = 10\log_{10}\frac{k(\pi\varphi)^2}{\lambda^2} \qquad (28)$$

Link budget and EIRP

$$P_{\text{Rx}} = P_{\text{Tx}} - L_{\text{f1}} + G_1 - L_{\text{fspl}} + G_2 - L_{\text{f2}} \qquad (29)$$

$$\text{Effective isotropic radiated power (EIRP)} = P_{\text{Tx}} - L_{\text{f1}} + G_1 \qquad (30)$$

Intermods

Three-station, third-order intermodulation frequencies:

$$2A + B - C$$
$$A + 2B - 2C$$
$$2A - B + 2C \qquad (31)$$
$$A - 2B + 2C$$
$$2B + C - 2A$$
$$2C + B - 2A$$

Noise

$$N_{\text{thermal}} = kTB \qquad (32)$$

Satellite link theory

$$N_{\text{o}}(\text{unit noise in 1 Hz of bandwidth}) = kT_{\text{sys}} \qquad (33)$$

In logarithmic terms,

$$N_{\text{o}} = -228.6 \text{ dBW} + 10\log T_{\text{sys}}$$

For receive signal power of C,

$$C/N_o = C/kT_{sys} \tag{34}$$

G/T – figure of merit

$$G/T = G\text{dB} - 10 \log T_{sys} \tag{35}$$

where G is the receiving system antenna gain.

$$T_{sys} = T_{ant} + T_{rx} \tag{36}$$

The revised satellite link budget equation

$$C/N_o = \text{EIRP} - L_{fspl} - \text{other losses} + G/T_{sys}(\text{dB/K}) - k(\text{dBW}) \tag{37}$$

Noise temperatures

$$T_{sys} = T_{ant} + T_{rec} \tag{38}$$

Receiver side of reference point, noise temperature of a cascaded network

$$\text{Tr} = \text{T1} + \frac{\text{T2}}{\text{G1}} + \frac{\text{T3}}{\text{G1G2}} + \cdots + \frac{\text{Tn}}{\text{G1G2}\cdots\text{Gn} - 1} \cdots \tag{39}$$

Noise temperature of a passive attenuator

$$T_e = T(L - 1) \tag{40}$$

where T is temperature and L is attenuation loss (not in log).

Transmitter side of the reference point

$$\text{Tant} = \frac{(L - 1)290 + T_{sky}}{L} \tag{41}$$

Reliability

$$\text{Availability}\,(A) = \frac{\text{Uptime}}{\text{Uptime} + \text{Downtime}} \tag{42a}$$

$$\text{Availability}\,(A) = \frac{\text{MTBF}}{\text{MTBF} + \text{MTTR}} \tag{42b}$$

$$\text{Unavailability} = 1 - \text{Availability}\,(A) \tag{42c}$$

Serial chain − add unavailabilities

$$A\ \text{system} = 1 - (U1 - U2 - \cdots - Um) \tag{42d}$$

Parallel chain − multiply unavailabilities

$$A\ \text{system} = 1 - (U1 \times U2 \times \cdots \times Um) \tag{42e}$$

Appendix 2

List of Symbols and Variables from Equations

v	Velocity	m/s
f	Frequency	Hz or 1/s
λ	Wavelength	m
E	Field strength	μV/m
S or Ψ or PFD	Power flux density	W/m^2
A_e	Effective antenna area	m^2
P_{Rx}	Receive power at output of antenna	W
P_{Tx}	Transmit power at input to antenna	W
L_{fspl}	Free space path loss	no units
d or D	Distance	km
h	Height (above mean sea level)	km
$f_{apparent}$	Apparent frequency	Hz
f_c	Carrier frequency	Hz
y	Aircraft ground speed	m/s
m	Modulation depth (analog)	no units
B_s	Symbol rate of the data signal	Hz
β	Modulation depth (digital)	no units
R	Bit rate	bits/s
r	Baud rate	symbols/s
M	Number of signalling states	no units
C	Channel capacity	bits/s
B	Bandwidth	Hz
S/N	Signal-to-noise ratio	no units
A	Traffic	erlangs
Y	Mean call arrival rate	erlangs/s
S	Mean call holding time	s

Aeronautical Radio Communication Systems and Networks   D. Stacey
© 2008 John Wiley & Sons, Ltd

I	Interfering power	W
K	Number of mobiles	no units
C	Wanted receive power	W
C/I	Wanted power to interference ratio	dB
Io	Interference power density	W/Hz
W	Signal bandwidth	Hz
Eb	Energy per bit	J
W/R	Spreading factor	Hz/bit
P	Power	W
G	Antenna gain	dBi, dBd
Φ	Antenna diameter	m
L	Loss	no units
N_{thermal}	Thermal noise power	W
T	Temperature	K
N_{o}	Noise spectral density	W/Hz
T_{e}	Effective noise temperature	K
T_{ant}	Antenna noise temperature	K
T_{sky}	Sky noise temperature	K
A	Availability	no units
U	Unavailability	no units
MTBF	Mean time between failures	usually hours
MTTR	Mean time to repair	usually hours

Appendix 3

List of Constants

k	Boltzmann's constant	1.38×10^{-23} J/K
c	Velocity of light in free space	3×10^{8} m/s
R	Radius of earth	6370 km
D_{geo}	Geostationary distance	35 784 NM
Z_{o}	Impedance of free space	377 Ω

Aeronautical Radio Communication Systems and Networks   D. Stacey
© 2008 John Wiley & Sons, Ltd

Appendix 4

Unit Conversions

Miles/kilometres	1.60934
Metres/feet	3.28084
Nautical miles to degrees	1/60

Aeronautical Radio Communication Systems and Networks D. Stacey
© 2008 John Wiley & Sons, Ltd

Appendix 5

List of Abbreviations

3G	Third Generation (pertaining to mobile network technology)
AC	Alternating Current
ACARS	Aeronautical Communication Addressing and Reporting System
ADS-B	Automatic Dependent Surveillance mode B
AFTN	Aeronautical Fixed Telecommunications Network
AIMS	Airplane Integrated Management System
AM	Amplitude Modulation
AM(R)S	Aeronautical Mobile (Route) Service
AM(R)SS	Aeronautical Mobile (Route) Satellite Service
AM(OR)S	Aeronautical Mobile (Off Route) Service
AM(OR)SS	Aeronautical Mobile (Off Route) Satellite Service
AMPS	Advanced Mobile Phone System
ANSP	Air Navigation Service Provider
AOC	Aeronautical Operation and Control
APC	Aeronautical Passenger Communications
APT	Asia Pacific Telecommunications (group)
ARINC	Aeronautical Radio Incorporated
ASK	Amplitude Shift Keying
ASMGCS	Advanced Surface Movement Guidance and Control System
ATC	Air Traffic Control
ATIS	Airport Terminal Information Service
ATM	Asynchronous Transfer Mode; *also* Air Traffic Management
ATN	Aeronautical Telecommunications Network
ATR	Aeronautical Transportation Racking; *also* Austin Trumbull Radio
AVLC	Aviation VHF Link Control
A/D	Analogue to Digital conversion

BER	Bit Error Rate
BFSK	Bipolar Frequency Shift Keying
BNC	connector type
BP	Basic Protocol
BVHF	Broadband VHF (Very High Frequency)
CAA	Civil Aviation Authority
CD	Compact Disk
CDMA	Code Division Multiple Access
CF	Cash Flow
CITEL	North and South America Telecommunications Group
CLIMAX	Trade name term for extended coverage
C-OFDM	Coded-Orthogonal Frequency Division Multiplexing
CP	Combined Protocol
CPDLC	Controller Pilot Data Link
CRC	Cyclic Redundancy Coding
CSMA	Carrier Sense Multiple Access (protocol)
D/A	Digital to Analogue conversion
DAMA	Demand Assigned Multiple Access
dB	Decibels, a logarithmic scale (see Section 2.3)
DC	Direct Current
DCE	Data Control Equipment
DCF	Discount Cash Flow
DCL	Departure Clearance
DECT	Digital Enhanced Cordless Telephone
DEMOD	Demodulator
DFSK	Differential Frequency Shift Keying
DL	Data Link
DME	Distance Measuring Equipment
DMR	Digital Microwave Radio
DPSK	Differential Phase Shift Keying
DSB	Double Side Band
DTE	Data Terminating Equipment
DWDM	Dense Wavelength Division Multiplexing
E1	2 Mb/s transmission rate
E2	8 Mb/s transmission rate
E3	34 Mb/s transmission rate
ECAC	European Civil Aviation Conference
EIRP	Effective Isotropically Radiated Power
E&M	Ear and Mouth
EMC	Electromagnetic Compatibility
EPLRS	Enhanced Position Location Reporting System
EPM	Electronic Programmable Mode
ES	Earth Station
ETSI	European Telecommunications Standards Institute
FAA	Federal Aviation Authority (of the United States of America)
F/B	Front to Back ratio
FDD	Frequency Domain Duplex

FDDI	Fibre Distributed Data Interface
FDM	Frequency Division Multiplexing
FDMA	Frequency Division Multiple Access
FEC	Forward Error Correction
FIS	Flight Information Service
FM	Frequency Modulation; *also* Frequency Management
FMG	Frequency Management (Group of ICAO)
FMS	Flight Management System
FOE	critical frequency, or highest frequency, that is returned from ionosphere
FOT	*See* MUF
FSK	Frequency Shift Keying
FSPL	Free Space Path Loss
FV	Future Value
GA	General Aviation
GBAS	Ground-Based Augmentation System
GFSK	Gaussian Frequency Shift Keying
GNSS	Geostationary Navigation Satellite System (or sometimes Service)
GPS	Global Positioning System
GSC	Global Signalling Channel
GSM	Groupe Speciale Mobile (Global System of Mobile Phones)
GSO	Geostationary Orbit
G/T	Gain/Noise Temperature (Figure of Merit for Satellite Transceiver)
HEO	High Earth Orbit (usually same as Geostationary Orbit)
HF	High Frequency (3–30 MHz)
HPF	Highest Possible Frequency
HVAC	Heating, Ventilation, Air Conditioning
ICAO	International Civil Aviation Organisation
IEJU	Initial Entry JTIDS Units
IF	Intermediate Frequency
IP	Internet Protocol
IRR	Internal Rate of Return
IS 95	a 3G mobile communications standard
IMT 2000	International Mobile Telecommunications
ISO	International Standards Organisation
ITU (UIT)	International Telecommunications Union
JAA	Joint Aviation Authorities
JTIDS	Joint Tactical Information Distribution System
k-factor	k is the Earth's effective radius multiplier
KVA	Kilovolt Amperes – unit of power (vectorial)
LAN	Local Area Network
LEO	Low Earth Orbit (satellite)
L/H	Left Hand
LNA	Low Noise Amplifier
LORAN	Long Range (Low Frequency Navigation System)
LOS	Line of Sight
LRU	Line Replaceable Unit
LSI	Large-scale Integration

MAN	Metropolitan Area Network
MCU	Modular Concept Unit
MEO	Medium Earth Orbit (satellite)
MES	Mobile Earth Station
MF	Medium Frequency (300 kHz – 3 MHz)
MFDT	Multifrequency Dial Tone
MIDS	Multifunction Information Distribution System
MLS	Microwave Landing System
MOD	Modulator
MOPS	Minimum Operating Performance Standard
MS	Mobile Station
MSK	Minimum Shift Keying
MTBF	Mean Time Between Failure
MTTR	Mean Time to Repair
MUF	Maximum Useable Frequency
MWARA	Major World Air Route Areas
NCS	Network Coordination Station
NF	Noise Figure
NM	Nautical Mile
NPG	Network Participation Group
NPV	Net Present Value
NTR	Network Time Reference
OCM	Oceanic Clearance Datalink Service
OFDM	Orthogonal Frequency Division Multiplexing
OOOI	Off OUT On and In (appertaining to scheduled flight stages)
OSI	Open System Interconnection
OWF	Optimum Working Frequency
PABX	Private Auxiliary Branch Exchange
PDH	Plesiochronous Digital Hierarchy
PFD	Power Flux Density
PIAC	Peak Instantaneous Air Count
PSK	Phase Shift Keying
P2DP	Packed Two Double Pulse
P2SP	Packed Two Single Pulse
P4SP	Packed Four Single Pulse
PTT	Public Telecommunication Operator, formerly Posts and Telegraphy Office
P34	a public mobile standard
QAM	Quadrature Amplitude Modulation
QOS	Quality of Service
QPSK	Quadrature Phase Shift Keying
RC	Resistor Capacitor (circuit)
RDARA	Regional Domestic Air Route Area
RF	Radio Frequency
RFI	Radio Frequency Interference
RH	Right Hand
RPE	Radiation Pattern Envelope
RR	Radio Regulations (part of ITU framework)

RTT	Round Trip Timing
SARP	Standard And Recommended Practices (ICAO documentation)
SC	Suppressed Carrier
SDH	Synchronous Digital Hierarchy
SELCAL	Selective Calling
SINCGARS	SINgle Channel Ground and Airborne Radio System
SNR	Signal to Noise ratio
SSB	Single Side Band
STDMA	Self-organizing Time Division Multiple Access
STDP	STandard Double Pulse
SV	State Vector
TACS	Total Access Communication System
TCM	Trellis Code Modulation
TDD	Time Domain Duplex
TDM	Time Division Multiplexing
TDMA	Time Division Multiple Access
TETRA	TErrestrial Trunked RAdio
TIS	Traffic Information Service
TMA	Terminal Manoeuvering Area
UAT	Universal Access Transceiver
UAV	Unmanned Aerial Vehicle
UHF	Ultrahigh Frequency (300–3000 MHz)
UST	Universal Standard Time
UWB	Ultrawide Band
VDL	VHF (Very High Frequency) Data Link
VHF	Very High Frequency (30–300 MHz)
VSAT	Very Small Aperture Terminal
VSWR	Voltage Standing Wave Ratio
WCDMA	Wideband Code Division Multiple Access
Wifi	*See* WLAN
WLAN	Wireless Local Area Network (sometimes called Wifi)
WRC	World Radio Conference
XPD	Cross Polar Discrimination

Index

A-BPSK 169
Abbreviations 339–334
Absorption 28
AC 242, 243, 267, 279
ACARS 2, 128–130, 132, 162, 210
Accidental (interference) 309
ACR 282, 283
Access schemes 65–71
ADS 4, 139, 202
ADS-B see ADS
AEEC 128
AFTN 187, 193
Air conditioning 243, 244
Air interface(JTIDs/MIDs) 149
Alignment 248
Allocation 123, 158, 171
Allotment 124, 158, 171, 179
Amplitude 14
Amplitude Modulation (AM) 39, 41
AM(OR)S 106, 145, 161–163, 165, 167, 168
AM(R)S 43, 106, 121, 145, 147, 161–163, 165, 167, 168, 182, 212
AMS 167
AMS(R)S 167, 169–171
AMSS 167
AM-ASK 129
AM-MSK 51
Analogue 39, 40
ANLE 208, 211
ANSP 192, 227
Antenna 10, 80, 81, 245–253, 294–300
Antenna coupler 300

Antenna layout (aircraft) 301–302
AOC 120, 121, 127, 131, 138
Application 212
A-QPSK 169
ARINC 129–131, 162, 273–275, 277, 278, 429, 629, 659
ARINC GLOBALINK 163
ASK 50–58
ATC 106, 108, 117, 120, 121, 123, 124, 127, 128, 132, 139, 140, 193, 248, 226
ATM (Asynchronous transfer mode) 191
ATN 187, 193
ATIS 124, 127
Availability 99, 113, 116, 160
Aviation VHF link control 132
Avionics 147, 259–304
Azimuth 81

Back plane 283, 284
Backward compatibility 110, 212
Band splitting 106
Bandwidth normalization 77
Base station 121
Bathtub curve 99
BER 57, 58, 118, 131, 137
Bipolar 52
BFSK 53, 54
BITE 277, 291
Blade (antenna) 300
Bluetooth 215
Boltzmann's constant 92
Break even point 222
Burn in 99

Burn out 99
BVHF 215

CAA 269
Cables 303
Cable leaks 309
Call set up delay 170
Call arrival rate (y) 64
Capacity 113, 212
Capital costs 221
Capture effect 49
Cashflow 224
Category 265, 267
Cavity filters 236, 237
C band 96
CDMA 67–71, 75, 92, 151, 201, 210, 214, 216, 217
Cell (radio) 120
Centrifugal 36
Centripetal 36
Certification 303
Channelization 140, 147, 183
Channel spacing 106
Circuit breakers 281
CLIMAX 50, 108, 118, 119, 120, 246, 286
Coaxial 238, 303
Coded OFDM 65, 66
COFDM 183
Coherent detector 43
Commercial aviation 270
Comm 2 tables 124, 322
Companding 44
Compression (see companding)
Congestion 307
Constants 335
Constellation diagram 57
Convergence 6, 208
Conversions 337
Convolutional coding 72, see also FEC, CRC
Cooling 283
Co-site 133, 247, 301
Costas loop 48
Coverage 113
CNS 6
CPDLC 130
CRC 62, 72
CSMA 131, 132
Critical Frequency 35

DAMA 67
Databus 273

D-ATIS 128
Datalink 2, 126–142, 215, 217, 218, 225
Data packing 153
DC 242–244, 267, 279
DCL 128
Decibel 14
DECT 215
Demodulation 158
Demodulator 46, 65, 94
DFSK 41, 54
Differential 52
Digital 40
Dipole 82
Direction finding 313
Disaster recovery 100
Discount cash flow DCF 225
Dish antenna (see parabolic antenna)
Distributed avionics 273
Diversity 74–77
DME 147, 156, 202
Doppler effect 37, 38
Down converter 46
DPSK 41, 55, 56, 206
D8PSK 56, 57, 130, 136
DSB-AM 41, 42, 105, 158, 217, 218, 235
DTMF 159
Ducts/Ducting (see refraction)
DWDM 191

E (electric field vector) 11
Earth station 241
Eclusian (distance) 59
Economics 221–227
Electrical disturbance 309
Electric field strength (E) 17
EIRP 174
Elevation 81
Effective aperture 16
EMC 93, 268, 307, 309, 310, 317
Encryption 127, 138, 151
Environment 229, 247, 259, 265, 267
EPLRS 146
Equalization 71
Equipment 229
Equipment racks 257
Equipment room 254
Erlang 64
ETDMA 201, 213
ETSI 240
Equations 325–332
EUROCAE 129, 304

Expanding (see companding)
Explosive atmospheres 265
Extended coverage 117
Eye diagram 32
E1 190
E&M 188

FAA 7
Fading 30, 32, 71, 74, 75
Fast fading (see selective fading)
FDD 70, 213, 215
FDDI 278, 279
FDM 63, 190, 191, 194
FDMA 66, 147, 150, 151
FEC 62, 72, 131, 138–139
Fibre optic 199
Figure of merit (see G/T)
Fluid susceptibility 266
FM 181
FMS 281
Folded dipole 82
Free space path loss (fspl) 15
Frequency coupling 117
Frequency diversity 75, 76
Frequency hopping 150
Frequency management 307, 322–324
Frequency modulation (FM) 39, 41, 49
Frequency shift keying (FSK) 39, 41, 53, 55
Fresnel 26
FSK 181
Fungus growth 266
Future communications systems 201–218
F/B ratio 84

Gain 80–87, 247
Gas absorption 29
Geo-stationary orbit satellite (GSS) 36
GES 170
GFSK 60, 139
Global signalling channel (GSC) 139
Gravitational 36
GNSS 147
Great Circle Distance 24
Ground wave 29
Ground installations 229–239
GPS 50, 210
Grade of service 171
Gravity 265
GSM 13
GSS 165

G703 190, 191
G/T 93, 172

H (magnetic filed vector) 11
Half duplex 44
Handover mechanism 133
Hardening (VHF) 124
Harmonics 88–92
HAVEQUICK 146
Header 153
HEO 168, also see GSS
HF 157, 158, 160, 162, 163, 165, 238, 241, 289
HF ACARS 164
HF datalink 162, 293
High earth orbit (HEO) 36
Highest possible frequency 35
High performance 254
Hilbert (modulator) 46
Horizontal expansion 110
Humidity 229, 230, 264
HVAC 244
Hysteresis 49

IATA 6, 226
ICAO 6, 14, 131, 134, 168, 211, 226,
IF 46, 245
IF combining 75, 76
IFF (identification friend or foe) 147
Immunity 268
Impedence 247
Inadvertent (interference) 309
Index PIN code 284
Indoor 230
Inflation 224
Infrastructure 229
Inmarsat 163
Inmarsat M 166
Inmarsat Swift broadband 203, 213
Interference 92, 307, 308
Interference to Noise I/N 311
Interleaving 72
Intermodulation 88–92, 124, 125–126, 246, 309
Internal rate of return 223
Internet protocol (IP, IPv4, IPv6) 156, 182, 191
Inter-symbol interference 32
Investment cost 221
Ionosphere 32
Ionosphere sounding 35
ISO 8202 132, 134
Isotropic 15, 81

ITU 6, 10, 121, 123, 124, 146, 159, 162, 177, 182
ITU-R (see ITU)

JAA 165, 269, 303
Jamming 310
Jam (resistant) 147
Jitter 152
JTIDS 145, 147

k-factor 22
Knife edge (diffraction) 26
Ku band 36

LAN 201, 208, 210, 214
Latency 37
L Band 36
Lightening 231
Link budget 87, 88, 93
Link 4A 11, 16, 147, 150
Link 2000+ 131
LNA 181
Log periodic antenna 84, 253
LOS (line of sight) 17, 30, 75, 115, 171
Low earth orbit (LEO) 36, 167, 168
LRU 282

MAN 208, 210, 214
Manchester code 207
Man made noise 92 (also see noise)
Macro-economics 226, 227
M-ARY 52, 53, 56, 58
Mast 246, 254
Master Minimum Equipment List 304
Maximum usable frequency 35
MCU 282, 283
Microwave 240
Microwave radio 194–196, 199
Message start opportunities 201, 202
MIDS 145, 147
Mean call holding time (s) 64
Medium earth orbit (MEO) 36, 168
Military (Aviation) 145, 271
Mobile satellite 165
Mobility 208, 212
Mode S 201, 205–207, 210
Modulation 38, 41–61, 158, 169
Modulation index (M) 42, 43
Modulator 44, 45
Mounting arrangements 248
MTBF 100, 235, 268
MTFD 188

MTTR 100, 235
Multipath 30, 72–75
Multiplexing 62, 199
MWARA 160, 163, 238

Natural noise 92 (also see noise)
Navigation 147
NET 147, 151
Net present value NPV 223
Network time reference(NTR) 152
Noise 92–98
NM (Nautical Miles) 10, 23
Noise figure 95
Noise temperature 93, 94
Non coherent detector 43
Normalization 77, 315
Notch (antenna) 300

OCM 128
OFDM 65, 66, 210
Omnidirectional 81
Optimization 248
Optimum working frequency 35
OSI 278
Outdoor 230, 245, 247, 257
Oxygen absorption 28

PABX 193
Parabolic 253
Parabolic antenna 86
Passenger communications 175
Passive receiver diversity 75
PDH 189, 191
Phase shift keying (PSK) 39, 41, 53, 55, 58, 162
PIAC 117, 120
Pirate 311
Point to point 240
Polarity 82
Polarization 11, 247
Power 14
Power dissipation 243
Power flux density (PFD) 16
Power supply 279
Pre-emption 171
Pressure 261–262
Pressurization 244
Priority 171
Private aviation 269
Propagation 10–16
Protocol 162, 169, 278
PSTN 167, 199

PTT 187, 194, 223
Pulse code modulation 180
Push to talk 44
P34 202, 213

QAM 41, 58, 59
Quality 113, 121
Quarter wave vertical antenna 82

Racking 282
Radio frequency (RF), 1, 10, 247
Radio regulations 307, 308
Rain zone 28
RAM 281
Receiver 10
Refraction 18
Reliability 99, 113, 211, 268
Reliability block diagram 102
Reliability (cost) 226
Resilience 211
RF environment 268
RMS (route mean squared) 45
Round trip timing(RTT) 153
RPE 81–87, 247
RS232 241, 242
RTCA 129, 262
Running cost 221

Safety case 225
Salt fog 266
Sand and dust 266
SARP's 113, 131, 134, 159, 206
Satellite 92, 163–175, 210, 213, 240, 300
S Band 4, 178
SC-Am 41, 48
SDH 191
Sector(ATC) 120
Security 211
SELCAL 159, 160
Selective fading 71, 72
Shannon's law 40, 60, 62
Shunt 300
Signal shaping 44
SINCGARS 146
Simplex 44
SITA 129–131
SI Units 10
Size 265
Sky noise 92
Sky wave 33
Smooth edge (diffraction) 27

Snell's law 31
Sniffer 239
SNR 46
S(N + I) 165
Software defined radio 217, 218
Solar flares 36
Space diversity 74, 76
Spectrum 178, 212
Spectrum management 307, 318–321
Splitter 239
Squelch 46
Squitters 35
SSB-AM 41, 46, 47, 158, 217, 218, 289–292
SSR 205, 206
Standardization 266
Sun spot 36
Surveillance 147
Survivability 147
Swamping (receiver) 124
Symbols 33–334
Synchronization 140, 152

Terrestrial backhaul 187–196
TCM 41, 58–60
TDD 70, 71, 202, 213, 215, 217
TDM 65, 189, 191
TDMA 134–137, 139, 147, 151, 201, 213
Technology 211, 212
Telecontrol 182
Telemetry 49, 177–186
Temperature 229, 261
Testing (for interference) 313
Thermal noise 92 (also see noise)
TMA 115, 216
Tower 231, 246
Transmission line 245
Transmitter 10
Trajectory 37
Trunking 62, 63
Two ray model 31

UAV 4, 177, 179, 182, 185, 186
UAT 128, 201–205, 210, 321
UHF 30, 145, 194, 208, 241
Unavailability 100
Up converter 46
UV exposure 230

Variables 333–334
VDL (see datalink)
VDL 0/A 128, 138, 141, 142

VDL1 128, 129, 141, 142
VDL 2 58, 128–130, 138, 140–142
VDL 3 58, 128, 129, 134–138, 140–142
VDL 4 61, 128, 129, 138–142
Vector 11
Vertical expansion 110
VHF 1, 12, 30, 32, 105, 124, 145, 147, 157, 160,
    162, 163, 167, 194, 208, 214, 216, 223, 232,
    233, 235, 241, 248
VHF datalink (see datalink)
Vibration 266
VOLMET 160
Voltage controlled oscillator (VCO) 49
Voting network 117, 118
VSAT 187, 197, 199, 241, 242
VSWR 82, 247, 292

Water absorption 28
WDN 191
Wear out 99
Weight 265
Weight loading 230

Whip (antenna) 295
White noise 92 (also see noise)
WiMAX (see 802.16)
Wind loading 230, 246
Wind speed 230
WRC process 123, 177, 185, 186, 210, 321
Wright Brothers 1

XPD 247
X25 132, 134

Yagi antenna 84

3G 2, 210
4G 210
4W E&M see E&M
5G 210
5/8 λ antenna 83
8.33 108, 218
802.xx derivatives 201, 207–209, 213
802.16 208, 209
802.17 209